3/28/91

The provident sea

The provident sea

D. H. CUSHING

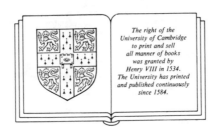

The right of the
University of Cambridge
to print and sell
all manner of books
was granted by
Henry VIII in 1534.
The University has printed
and published continuously
since 1584.

CAMBRIDGE UNIVERSITY PRESS

CAMBRIDGE   NEW YORK   NEW ROCHELLE

MELBOURNE   SYDNEY

Published by the Press Syndicate of the University of Cambridge
The Pitt Building, Trumpington Street, Cambridge CB2 1RP
32 East 57th Street, New York, NY 10022, USA
10 Stamford Road, Oakleigh, Melbourne 3166, Australia

First published 1988

Printed in Great Britain at the University Press, Cambridge

British Library cataloguing in publication data
Cushing, D
The provident sea
1. Fishing management
I. Title
639′.2 SH328

Library of Congress cataloguing in publication data
Cushing, D. H.
The provident sea / D. H. Cushing.
p.   cm.
Bibliography: p.
Includes index.
ISBN 0 521 25727 1
1. Fisheries – History. 2. Fish trade – History. I. Title.
SH211.C87 1988   87–33029
333.95′6′09–dc 19   CIP

ISBN 0 521 25727 1

CONTENTS

PREFACE

In earlier ages conflicts between nations over fish were settled by war. In the first decades of the present century, in Europe and North America, efforts were made to resolve such conflicts with scientific evidence. The point was that it was published and presented for decision.

The present industrialized fisheries are contrasted here with earlier ones before the industrial revolution. The older fisheries for cod and herring persisted for centuries with only minor changes in catch abundance, so far as can be seen with somewhat limited evidence. They were probably secure until capture was mechanized. The industrialization of capture, with its greater efficiency, reduced the stocks of fish much more than in earlier ages. The first industrialization in the North sea and in the Alaska gyral reduced stocks and raised the problem of whether action could be taken world wide. The international fishery institutions were established in the first decades of this century and became effective in the thirties and forties. The second spread of industrialization across the world ocean reduced most of the stocks and exposed the weakness of the then international commissions to which nations had adhered voluntarily. Further, fleets of trawlers appeared off the coasts of many countries bordering the tropical and subtropical seas. The consequence was the Law of the Sea Conference, and by 1977 the power to regulate fisheries was effectively vested in the coastal state.

The history of management is described as it took place, with the scientific material as it was originally published. There exists an extensive literature on how fish stocks should be managed, but this has not been cited because only those parts which have been translated into action have been mentioned here. Further the study is really limited to the North Atlantic and the North American coast of the Pacific. Much has happened elsewhere but the history is often intermittent or very recent.

Thanks are due to Stephen Charlton, who made the line drawings. The chapters on whales and seals were read by Dr J. A. Gulland FRS; that on Fisheries Research since 1965 was read by Dr John Shepherd,

Professor Brian Rothschild and Dr M. P. Sissenwine; that on the institutions since 1977 was read by Professor Rothschild and Dr Sissenwine. I am grateful to each for their criticisms but of course the errors remain my own.

I am also grateful to Peter Jenkins of Lowestoft, who allowed me to use the picture reproduced on the cover, and to Mrs Sandi Irvine who edited the text.

198 Yarmouth Rd, Lowestoft, Suffolk   D.H.C.
January 1987

UNITS USED IN THE TEXT

For purposes of comparison and updating, most of the units from the original sources have been converted, where necessary, to metric. This has meant that some approximations have been introduced (e.g. 1 fathom $\simeq$ 2 m), but exact conversions, or in some instances a conversion factor, have been given where appropriate.

Where a legal limit is mentioned, the original defined units are used, e.g. 200 mile limit.

Apart from the usual unit conversion factors, the following should be noted:

US gallon = 3.7853 l
British gallon = 4.5460 l
US bushel = 35.2381 l
British bushel = 36.37 l
Tun = 100 kg
Barrel $\simeq$ 0.125 tonne (herring)
    $\simeq$ 0.167 tonne (whales)
British (imperial) ton = US long ton = 1.016 tonnes
US short ton = 0.907 tonne
Quintal = 98.39 ($\simeq$ 100 kg)
Quintal of dry fish $\simeq$ 500 lb (227 kg) of fish from the sea (after heading, splitting and drying)
(The conversion of a quintal of dried fish to tonnes is $(500/112.20)0.984$ = 0.22. That quoted on p. 73 is 0.23.)

For shipping, the US long ton or imperial ton was used; for cargo vessels 'dead weight' is 'displacement' minus 'light weight'.

1

Fisheries in prehistory and in antiquity

The relics of the past lie in the earth and in documents. They include accounts of fishing as sport, or as industry and the remains of fishing gear. Necessarily they are incomplete as compared with information collected today. The evidence, however, is positive and can be used with care. Before the material is presented, the four standard gears in use today throughout the world will be described briefly (many others will be described in more detail in later chapters), and this will help to illustrate the material from the more distant past.

The four gears

The simplest gear is a hook on the end of a line; today hooks vary widely in shape and in material and lines are usually made of synthetic fibres, which have replaced the original flax and hemp. Long lines, with hooks every fathom or so, are laid on the sea bed and the great lines used off Iceland and Greenland extended for many kilometres. Each hook in such a system has to be baited as it flies from the drum, perhaps with mussels or herring. A larger system is the pelagic long line used for catching tuna-like fishes in the subtropical ocean; the barbs are bare, the lines are suspended from floats at the surface and the whole system is up to 80 km in length (Figure 1(*a*)).

A drift net is floated up by buoys from a heavy messenger rope attached to a vessel, a drifter. The vessel and the long curtain of nets drifts, or drives with the tide. The system must drag a little in the sea, and fish migrating with the tide swim into it and are meshed by their gill covers. In the old East Anglian herring fishery in the southern North Sea, about seventy nets were used, a curtain of up to 2.5 km in length. Such nets are now costly, and if heavy with fish they need a crew of about ten men to haul them for many hours. The set net is anchored to the sea bed and herring swim into them on their spawning grounds; the method is quite distinct from that of the drifter (Figure 1(*b*)), which drives with the tide.

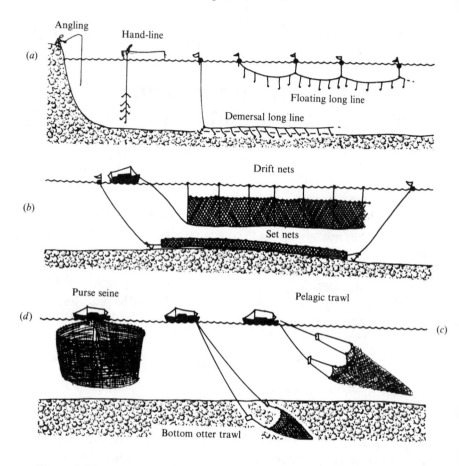

Figure 1. The main types of fishing gear, seines, trawls, gill nets and lines, all of which exist today; traps are illustrated in chapter 2 (from Cushing 1975).

(a) Lines: hooks are attached to lines to catch fish; they may be hand lines, demersal long lines a few kilometres in length or pelagic lines up to 80 km in length (Cushing 1975).

(b) Gill nets: long curtains of netting into which the fish swim and become caught by their gills. A drift net, perhaps 2.5 km in length is suspended from the surface or floated up from a heavy rope attached to the vessel; the whole system drifts or drives with the tide. Set nets are anchored to the sea bed.

(c) Trawl: a conical net with a broad low mouth dragged along the sea bed. The otter trawl is kept spread with doors or otter boards and the headline is lifted with spherical floats. In a beam trawl, the same functions are carried out by a beam fastened to iron trawl heads. A pelagic trawl is a large net with a roughly square mouth, towed in midwater.

(d) Seines: a purse seine is an enclosing net shot around a shoal, which is closed or pursed below by ropes from the vessel, laced through rings at the bottom of the net. With a Danish seine, a circle of net laid on the sea bed is hauled from the vessel and herds the fish into the net.

There are many varieties of trawl in use today developed from the dredges and beam trawls of earlier times. The beam trawl was that used by the North Sea smacks in the nineteenth century; the beam was attached to triangular trawl heads. The net was a long conical bag spread back from the beam to the cod end, which collected the fish as the trawl was towed through the sea on a single warp; advanced forms of this trawl are used today in the southern North Sea. The otter trawl is spread by 'doors' on two warps which are handled by a double-barrelled winch on the trawler's deck. Most trawls are worked on the seabed for bottom-living or demersal fish, but there are also trawls that are operated in the midwater for pelagic fish (Figure 1(c)).

The seines are encircling nets, shot from the shore or out in the open sea. A curtain of net is shot in a circle until the ring is closed at the surface and is 'pursed' below. Such nets are today very large and hundreds of tons of herring-like or sardine-like fishes can be taken with them. The Danish seine works in an analogous way on the sea bed, where the ring is closed by ropes laid on the bottom (Figure 1(d)).

Around the seas of the world, traps, pots and creels are used; fish and crustaceans swim into them to take the bait provided. Salmon and tuna are taken in corridors of netting, the salmon in fixed engines in British waters and the tuna in *madragues* in the Mediterranean.

The prehistoric evidence (Rau 1884)

Hooks, barbed harpoon heads and gorges have been found in palaeolithic remains at Kesslerloch (near Zürich). Bones of salmon, trout, pike, bream, white bream, dace and chub have been found in the valleys of the Dordogne and the Vézère in France. Drawings of carp, pike, eel and spurdogs on reindeer horn have been recovered. Compound fish hooks like those used by the North American Indians to catch the halibut off Cape Flattery were taken from Kesslerloch.

Neolithic remains of oyster, cockle, mussel, periwinkle, herring, cod, dab and eel were found in Danish middens; remnants of salmon, pike, perch, carp, dace, chub, turbot and rudd have been taken from the Swiss lake dwellings. From the same areas have been recovered bone fish hooks, bark floats, grapnels for recovering lines, deep horn harpoon heads and fragments of nets made of flax with sinkers. Dug out boats were found with anchor stones and what appeared to be netsmen's needles. From the bronze age, hooks of different shapes were recovered from the Lake of Neuchâtel; the Romanshorn was a large 15 cm hook. At Cudrefin (near the lake), a well-shaped boat was found, made of oak, 7–10 cm thick with cross ribs. Lhote (1959) published reproductions of cave drawings from the Sahara which show a fisherman working hook and line from a canoe (*c.* 3500 BC).

From the prehistoric evidence we conclude that some of the fish we eat today were taken by hook and line long ago, with nets from the banks of rivers, from the shores of seas and from boats. In other words the fisheries in the dawn of history on rivers and close to the sea shore may not have differed very much from those which we know today in the same places.

Fisheries in ancient Egypt

Radcliffe (1921 and references therein) gives an account of the fisheries of the Old, Middle and New Kingdoms. Spears, harpoons and bidents (which are still in use in Lebanon and Syria today) with long lines for recovery were used from river banks or papyrus punts in about 2000 BC. Copper hooks, both barbed and barbless, of 2–6 cm in length were found in tombs of the first dynasty, the earliest period. Nets appear as bags in remains dating from the third and fourth dynasties of the Old Kingdom. In the rock tomb of Deir el Gebrawi, there is a picture of seven fishermen hauling a seine from a boat, with eight species of fish in the catch. Needles are shown in all ages, with spindles; handnets, double handnets, cast nets, stakes and seines were used. The seines were weighted down with stones. Nets were made with flax in double and triple strands with meshes of from 0.3 cm to 1.2 and 1.9 cm. Weels (or wicker fish traps) were small (1.5 m) or very large (3–4 m, for several men).

Certain fishes were venerated; *Lepidotus* and *Phagrus* are recorded as such by Herodotus, and Strabo noted a respect for the Nile perch. Parts of the god Osiris were said to be eaten by the shark (*Oxyrhynchus*). In Ptolemaic times, the catch was taxed up to one quarter of its value; there was also a tax on the right to fish, particularly in waters owned by the temples. The nets in the marsh country were 'engines of encirclement' or purse-seine-type nets; the fisherman used it by day to catch fish and by night as a bedspread. Fish were dressed on the boat and dispatched quickly to market; indeed they were exported to Palestine in baskets or in barrels.

Assyrian fisheries

Handlines and nets were used from light boats, according to Radcliffe (1921), but there appear to be no records of spears or rods. Creels were used and a 'double knotted surrounding net'. As long ago as 2500 BC, there was a fish pond in each town, under the charge of a keeper. Sea fish – turbot, sole, swordfish, shark, flying fish and sturgeon – were kept in *vivaria*.

Jewish fisheries

As in Assyria, rods were absent. But handlines, spears, bidents and cast nets were certainly used. Scaleless fish were initially excluded by law (siluridae, skates, lampreys, eels and shellfish); weirs and fences were forbidden, and nets were made in the main worked from river banks. Fish ponds were a late development (Radcliffe 1921).

Fisheries in classical antiquity

One of the difficulties in examining the classical texts is that the names of fishes do not always correspond with those of today; I have used d'Arcy Thompson's (1947) *A glossary of Greek fishes* (and references therein). Aristotle refers to 110 species in his *Natural history*, but only fifty can be identified, of which all save six came from the sea. Homer (quoted by Radcliffe 1921) refers to spears, a net (a beach seine) of 'all ensnaring flax', rods and hooks and lines. Theocritus in the *Fishermen's dream* refers to the instruments of their toilsome hands, the fishing creels, the rods of reed, the hooks, the sails bedraggled with sea spoil, the lines, the weels, the lobster pots woven of rushes and the seines. Leonidas wrote of a well-bent hook, a long rod and a line of horse hair.

Aelian (AD 170–230; Hercher 1971) recorded four gears – nets, spears, weels and hooks; artificial flies made of red wool round a hook were bound with two feathers from under a cock's wattle. He wrote that grayling can be caught only with a hook baited with a particular gnat. The method of catching young tuna is obviously a form of trolling, lines with hooks trailed from the stern of the fishing vessel, with bait wrapped in red wool and gull feathers. Ausonius (*c.* AD 310–393; Hosius 1967) wrote of fishing, perhaps for salmon or trout, in the Moselle. Men in boats drag nets in mid stream, watch the corks of little nets in shallower water and on rocks, anglers with rods scan the floats bobbing in the water. He also referred to a knotty seine and to drag nets buoyed on their cork floats.

Nets were of all sorts and kinds in shape, make and size. Alciphron wrote that 'scarce a fathom in the harbour of Ephesus but held a net'; once the sole haul was the putrid carcass of a camel, so the net was fairly large. Rods were jointed. Lines were usually woven of horsehair, flax and broom; hooks were made of iron or of hard bronze (i.e. of tin and copper, not zinc and copper) and there were one or two sharp barbs. Lines were used with floating cork and lead attached close to the hook; sliding corks were used to regulate the position of the lure. Myrrh dissolved in wine was used on the bait to intoxicate and cyclamen was used with bread to poison. Istrian fishermen caught silurus (*Parasilurus aristotelis* Agassiz) of about 180 kg and hauled it in with oxen (Radcliffe 1921).

Probably the fullest single account of Roman fishing is given in the Halieutica of Oppian (1928) (*c.* AD 170). He lists the fishes that can now be identified: red mullet, *Trachurus*, sole, mormyrids, mackerel, carp, blenny, grey mullet, bass, conger, sea horse, gurnard, perch, rainbow wrasse, *Sciaena*, dory, parrot wrasse, *Muraena, Serranus gigas, Glaucus, Dentex, Scorpaena, Sphyraena, Balistes,* tuna, swordfish, pilot fish (*Naucrates*), pilchard, shad, sucker, spiny crayfish, lobster, crab, prawn, hermit crab, octopus, cuttlefish, oysters, sea urchins, mussels, razor shell, nautilus, sawfish, dogfish, spiny dogfish, spotted dogfish, dolphin, whale, seal and turtle, amongst others.

Some delight in hooks and, of these, some fish with a well twisted line of horse hair fastened to long reeds, others simply cast a flaxen cord attached to their hands, another rejoices in leaded lines or lines with many hooks. Others prefer to array nets; and of these there are those called casting nets, and those called draw-nets, drag nets and round bag nets and seines. Others they call cover nets, and with the seines, there are those called ground nets and ball nets and the crooked trawl; innumerable are the various sorts of such crafty bosomed nets. Others again have their minds set rather upon weels which bring joy to their masters while they sleep. Others with the long pronged trident wound the fish from the land or from the ship, as they will (Oppian; see Mair 1928).

Some sharks bit through the lines, grey mullet leapt over the cork lines, the moray eel circled for a wide mesh to slip through, a hooked bass enlarged the wound and escaped and deep sea fishes were landed with onions or with bare hooks. There was a large variety of odours used in the lures; broiled octopus or crayfish were said to raise bream; parched vetches moistened with fragrant wine yielded shad, pilchard and horse mackerel; red mullet were said to like smelly baits and grey mullet a floury bait with mint. Mackerel rushed into the (drift) nets, some landed in the wider meshes and leapt out, but others, penned in the narrow openings suffered a bitter fate by strangling. At three places, off Sicily, off the mouth of the Rhone and off the Spanish coast, the men fishing for tuna found places neither open to the wind nor 'straitened under beetling banks'; a watcher from a hill told his comrades of the sizes, shapes and direction of shoals and the tuna poured into the *madragues*.

Oppian (see Mair 1928) describes the capture of whales with a line of many strands of well-woven cords, as thick as the forestay of a ship. The well wrought hook was rough and sharp with barbs, alternately on either side. A coiled chain was cast about the butt of the hook, a stout chain of beaten bronze. Well-benched ships were used with quiet oars. The hook was baited with liver. The whale took the hook and dived, whereupon the fishermen let large bladders go, fastened to the line. When the bladder rose to the surface the whale was tiring. When he surfaced he was attacked with strong harpoons, stout tridents, bills and axes and 'with the end of his tail, he ploughs up the waves of the deep'. The whale was towed ashore by its teeth. Fin whales and sperm whales live in the Mediterranean (Viale 1981) as do pilot whales; perhaps the ancient Greeks caught pilot whales.

Oppian is very interesting on the capture of dolphins: 'the hunting of dolphins is immoral . . . for equally with human slaughter the gods abhor the deathly doom of the monarchs of the deep; for like thoughts with men have the attendants of the god of the blooming sea; wherefore they practise love of their offspring and are friendly one to another.' I shall return to this succinct statement in a later chapter on whaling. Off Euboea, fishermen took their catches under 'the swift gleam of the brazen lantern'; the dolphins chased the fish towards the 'well pronged tridents' and subsequently came to ask for their share.

Homer (see Radcliffe 1921) does not refer to fish in banquets, but the Romans did. They took fish from the Tiber, the Po, the Danube, the Rhine and from the north Italian lakes, but they preferred sea fish; in Diocletian's edict of AD 301, the best-quality sea fish were considered to be twice as valuable as the best-quality freshwater fish. The most valuable fish were red mullets, sturgeon and turbot and the prices of red mullet were very high indeed. The Roman nobles spent large quantities of money on aquaria (Radcliffe 1921).

Fish were used in sacrifices and formed the basis of a trade in salted and pickled fish (indeed salsamentum was also used to prevent scurvy); Olbia in Sardinia was an important market for this trade. Archimedes (see Radcliffe 1921) refers to an aquarium of 95 470 l on board ship (part of the corn traffic between Sicily and Egypt), which was lead lined.

Roman law illuminates the nature of the fisheries. Fish and wild animals were among *res nullius*, things belonging to no one. They become the property of the person who first 'reduces them to their possession'. The seas and public rivers could not be owned by individuals and so no individual could be prevented from fishing in the sea and in such rivers; however, a cove alongside private property might be enclosed by stakes and a backwater of a public river could be acquired by prescription. Such a structure is very like that in British waters in the nineteenth and early twentieth centuries, which implies that the fisheries were fairly well developed (Radcliffe 1921).

Smith (1876) discussed the distribution of fisheries in so far as they were known. The Black Sea and the Sea of Azov were the most important sources of fish. Fishermen's societies worked large vessels which sailed to Spain, Portugal and North Africa. They used lines, nets, seines and harpoons; hooks were made of copper, or iron covered with tin and iron chains were used to catch sharks. Tuna were taken from the Straits of the Bosphorus, Italy, Sicily, Sardinia, the Straits of Bonifacio, the Straits of Messina, the Straits of Cadiz (*sic*), off Elba, off France and off Spain. Mullets were taken from the Straits of Gibraltar. Salted tuna came from Euboea, Icaria, Cefalu, Samos, Icaria, Orbetello, San Stefano and Malaga (Malach means 'to salt'). There is argument about *Asellus*, a gadoid, which might be hake or a smaller one; whatever it was it was highly valued.

According to d'Arcy Thompson (1947) turbot was caught mainly in the Black Sea, but it was excellent off Ravenna (the northern Adriatic is a relict sea). *Oreochromis nilotica* was taken in the Nile. The sturgeon was celebrated in Rhodes; it was a costly luxury in Rome. It was also caught at holes in the ice in some lakes, as it is today in Lake Winnebago in the State of Wisconsin. The bonito was gregarious and migratory and spent its summer in the Black Sea; it was best in Istanbul. Dolphins were regarded as a grievous enemy of lesser fishes and to hunt them was sinful and displeasing to the gods; they were, however, taken by the Thracians and off Pharnacia in Chaldea, where they chased the tuna and were caught for their blubber. Eels were kept over winter in clean water in small tanks; they were caught in great numbers where the Mincius flows from Lake Garda. Tame eels were found in the fountain of Arethusa at Chalcis in Euboea. Tuna were taken by a seine net paid out by six boats at the entrance to the Gulf of Argolis. The 'hooers' (the Cornish name) or *speculatori* on tall masts or high cliffs had to be loud voiced, sharp sighted and quick at figures. Bass provided excellent food 'between the bridges' in Rome. The Nile perch dried and salted was considered the prize fish, except for *Tilapia*. Swordfish were caught of Scilla as they are today from swordfish-shaped boats. Bluefin tuna were taken with great iron hooks on strong ropes; when hooked they try to enlarge the wound in order to escape. They migrate past the Pillars of Hercules and were caught plentifully in the Iberian and Tyrrhenian Seas. Young Tuna (or 'pelamys') gathered at the entrance to the Sea of Azov; they were caught by an engine armed with hooks and weighted with lead, which was dropped on the fish as they lay in deep water. Aelian (see Hercher 1971) wrote of swift boats with ten rowers and of hooks baited with a coloured rag and trimmed with a gull's feather 'so as to flicker gently at the surface of the sea'; this is one method used today for mackerel in the waters off western Europe and off North America. Oysters were grown on hurdles and were transplanted to Chios from Ryrrha in Lesbos, but 'maiora Lucrinis, dulciora Britannicis, suaviora Medulis, acriora Ephesiis, pleniora Lucrensibus, sicciora Coryphantenis, teneriora Histricis, candidiora Circeiensibus . . . rufa Hispaniae, fusca Illyrico, nigra et carne et testa circeiis.' d'Arcy Thompson's evidence is most valuable because he was a fisheries biologist and at the same time a notable classical scholar; hence his survey is of some importance.

Arrianus (Iliffe Robson 1929–30) refers to the ichthyophagi in the Arabian Gulf. They ate nothing but fish, ground to a meal for themselves and their cattle; they were said to be dressed in fish-skin clothes and to live in huts made of whale skeletons. Perhaps, in fact, they lived in the Hadramaut in Southern Arabia on the shore of a notable upwelling area. They had nets which could cover two stadia (0.4 km) and which were made of the inner bark of palm trees. The *dictymum* was a cast net, the *amphiblestoon* was a seine (perhaps a shore seine); *sagenai* resembled modern seines which stretched many roods to seaward; *sphaerenoci* were pockets in a seine; *hypochaei* were small round nets and *gangamai* were drag nets or dredges; *gryphai* and *kurtai* were traps of bent osier twigs; *panagreas* were nets, generally.

Conclusion

Most classical writers, apart from Oppian, were not interested in fisheries as such, and many modern translators are concerned with angling. Yet from their material and the prehistoric evidence emerges a picture of fisheries in the distant past which is quite compatible with accounts today. The four gears were there, the same fish were taken as can be found in any Mediterranean restaurant today, there was trade across considerable distances and there was a legal basis to the practices of fishermen.

2

The preindustrial fisheries

By 1987 it has become difficult to describe the fisheries before industrialization. However, some historians and some observers give a reasonable picture, if only a qualitative one. The historians are Parona (1919), who described the tuna fishery in the Mediterranean in the nineteenth century and Cadoret *et al.* (1978), who gave an account of the Breton fisheries for sardines and albacore in the nineteenth and twentieth centuries. The observers were: Duhamel du Monceau (1769), who described gear and fisheries in France in the eighteenth century; Brown Goode (1879, 1887*a*), who wrote histories of various fisheries of the coasts of the United States; and Hornell (1905, 1914, 1916*a*, *b*, 1917, 1925, 1938, 1950), who was a fisheries administrator in India from the first decade of the present century until the Second World War.

Fishing gear as illustrated by Duhamel du Monceau (1769)

There are two main groups of gear, those based from shore and those worked mainly at sea. The two groups overlap to some degree.

Gear used from shore

The simplest nets are the cast net and lift net (Figure 2). Both can be seen today in many parts of the world. Cast nets are thrown over small shoals of little fish in shallow water. Shrimps, prawns and crayfish are caught with lift nets on the rising tide. Figure 3 illustrates the familiar crab and lobster pots used off rocks and from small boats throughout the world. Figure 4 shows traps left at low tide with fish retained in them.

Gear used at sea

Lines

Fish caught with hook and line met much of the preindustrial needs. Figure 5(a) shows an array of lines laid on one low tide for capture on the next. In Figure 5(b) is shown a long line on the sea bed, marked by a buoy at the surface. Drifting lines are illustrated in Figure 5(c); the boat is anchored and the lines drift in the tide in rather shallow water to catch fish such as whiting.

Gill nets, trammels and drift nets

Drift nets comprise a long curtain of net suspended in the water by floats near the surface and weights on the lower edge. The whole system drifts or drives in the tide, attached to the boat; fish such as herring, sardine or mackerel swim into the meshes because the curtain of netting drags a little in the water (Figure 6(a)).

A gill net is like a drift net but is anchored to the seabed. It can be used on the foreshore between tides or in somewhat deeper water. Fish are carried into it on the tide (Figure 6(b)). The trammel net resembles the gill net but consists of up to three superimposed nets of different mesh sizes, which can catches fishes of all sizes.

Seines

The simplest form of seine is the beach seine, which is like a drift net or gill net dragged by both ends towards the shore (Figure 7(a) and (b)). The same operation can be carried out at sea (Figure 7(c)), although the system does not seem to have the lines needed to purse the seine. A more remarkable equipment is the form of seine hauled by ropes on the sea bed, as does the Danish seine today; the boat is held by a sea anchor (Figure 7(d)).

Trawls

A trawl is a conical bag dragged across the sea bed to catch fish on the bottom. The trawl illustrated in Figure 8 was used on the coasts of Brittany and Poitou. It was 16 m long, and hauls of four to eight hours were made in winter and two hours in summer. At the end of the haul, the sail was lowered and the trawl was hauled by hand. A similar gear called the *gangui* was used off Provence.

Figure 2. (*a*) The cast net; (*b*) the lift net, and (*c*) lift net from a scaffold (Duhamel du Monceau 1769).

Figure 3. Crab and lobster pots (Duhamel du Monceau 1769).

Figure 4. Fish traps at low tide (Duhamel du Monceau 1769).

Some Indian fishing gear described by Hornell

Hornell (1925, 1938) described the fishing methods of the Madras Presidency, on the Coromandel coast (east) and on the Malabar coast (west). On the Coromandel coast lift nets were used from catamarans for pomfrets, mackerel, sardines and carangids. The Thuri Valai was a small cotton trawl with a hempen cod end towed by two catamarans, rowing. The Kola Valai was a boat seine of 6 mm mesh, with long wings and the mouth kept open with poles; it was hauled quickly for garpikes and halfbeaks by two catamarans. Shore seines and gill nets were used up and down the coast. The original drift nets were 400 m long and were made of thin hemp, armed with floats but no sinkers; a cotton drift net was introduced in 1875. One of the most interesting fisheries was that for flying fishes from large catamarans (9–11 m in length with two lateen sails) 19–32 km offshore; they streamed bundles of the leaves on which the flying fish lay their eggs. The fish were caught in the filamentous glutin of the egg masses.

In the backwaters of the Malabar coast there was an array of fish spears, harpoons, bows, crossbows and blow guns. Basket traps, filter traps, screen barriers and plunge baskets were commonplace; beach seines, stake nets (an anchored net which fishes in the tide) and dip nets (Figure 2(b)) were also used. At sea, long lines of 400–1200 m in length were used to take catfish, dogfish and rays, amongst others.

Hornell (1950) described the fishery for the Bombay duck (*Harpodon neherens* Hamilton Buchanan) in the Gulf of Cambay. The 20-tonne boats with a crew of seven to nine were fitted with a single mast and a lateen sail. Each man brought two bag nets with ropes, buoys and floats. The 46 m long bag net was secured by four short ropes to two large posts driven into the sea bed. It was a tidal fishery; the Bombay duck came in large shoals and large catches were made. The fish were sun dried and packed into large flat circular bundles, each containing about 2000 fishes.

The pearl and chank industries of India

The main fisheries for pearl oysters and the Indian conch or chank were concentrated in the Gulf of Mannar, north-east of Cape Comorin, and off the north-east coast of Sri Lanka. Both animals live on distinct shallow banks, characterized as pearl or chank beds. Both had religious or social significance and the fisheries are very old.

The pearl fisheries (Herdman 1904 and references therein)

The pearl fisheries in the Gulf of Mannar were referred to by Pliny and Ptolemy speaks of Komar (Cape Comorin) and the Kareoi or Karaiyar the coast people in the region of Tuticorin. Periplus of the Erythrean Sea in the first century AD describes the geography of the pearl fishery. But, according to the Mahawanso, King Vijaya gave pearls to his father-in-law in 550–540 BC; in 306 BC, King Devanampiyatissa sent eight kinds of pearls on an embassy to India.

In 1330, Friar Jordanns recorded that 8000 boats worked the pearl fisheries of Ceylon and Tinnevelly; the centre was Kolkhoi (Kayal referred to by Marco Polo), a port not far from Tuticorin, now silted up. The fishermen were Parawas; they dived for pearls, lived ashore and were paid annually by the rajas. Merchants came to the region from all parts of

India and judges were appointed to settle disputes. The headmen of the Parawas became rich and powerful. When the Moors spread across India they dived from Tuticorin; the Parawas killed thousands of them.

When the Portuguese reached India in 1470, the Parawas became Christian. In 1528, Manuel de Gama cleared the coast of robbers. The fishery started in the beginning of March when 4000–5000 boats put to sea. A merchant would fund four to six boats. Each boat had a master and eight or nine divers and would anchor in 10–14 m. As a sampling venture, three boats would go out to about 5 km from shore and each would bring back a thousand oysters, which were opened in the presence of the merchants. Each merchant constructed an enclosure of stakes and thorns, leaving a narrow passage for an entrance.

The boats sailed before dawn; each carried 60 lb stones attached to it by strong ropes. The diver was taken down by his stone; a second rope

Figure 5. Line fishing: (*a*) an array of lines laid at low tide; (*b*) a long line laid on the sea bed; (*c*) a drifting long line to catch cod-like fishes (Duhamel du Monceau 1769).

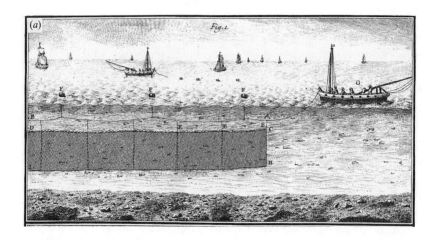

Figure 6. (*a*) A drift net; (*b*) a set net (as (*a*) but anchored to the sea bed) (Duhamel du Monceau 1769).

around his body (with a small bag) was held by two sailors on the boat. He stayed down for two credos (or the time to say the creed twice over), signalled on his rope and was hauled up. A second diver went down as soon as the first surfaced. The diver kept pearls from oysters opened underwater. Children stole oysters ashore.

The fishery lasted from 11 March to 20 April. The fair went on for fifty days, and during the last nine days the enclosures were cleaned. On the last day of April the merchants met to share the pearls and group them into nine classes; they were exported to all parts of the then known world. Amber and coral, both black and red, were washed ashore and also exported. By 1622, the pearl fishery had declined. The nayak (or local ruler) granted seven free boats, with 13¾ stones to each boat.

In 1658, the fishery and the port of Tuticorin were transferred to the Netherlands East Indies Company. The fishery was profitable in 1663, and by 1697 it was considered 'an extraordinary source of revenue'. By 1700, ten or a dozen vessels would test the banks and sample a few thousand oysters. If a fishery was to take place, traders and shipping assembled. The fishery was started by a cannon shot and the boats put to sea preceded by two large Dutch sloops. When a boat reached its marked place the divers went down with their stones on a rope round their waists, but secured to the boat; the oysters were put into a small sack. On a signal, the diver was hauled to the surface and a second sent down. In other words the method had not changed since the time of the Portuguese. The strongest men could not dive more than seven or eight times a day. The pearls were sorted into colanders and the Dutch bought the finest. By 1746, the free stones had been abolished and the fishery languished (Hornell 1905).

In 1796, the fishery was taken over by the English East India Company. The positions of the Indian fishery are shown in Figure 9 (Hornell 1916*b*). The Ceylon fishery took place in three main banks on the north-east coast of Sri Lanka, of which only the Cheval Paar remained productive at the turn of the present century (Herdman, 1904). Figure 10 shows the yield in thousands of rupees from the two fisheries in India and Sri Lanka (1879–1912; and in numbers from Sri Lanka (1800–1930). During the nineteenth century, the Sri Lankan fishery predominated, yielding over half a million rupees in some years. The Indian fishery, however, then yielded only about 100000 rupees (Hornell 1925).

Hornell (1916*a*) was impressed by the irregular nature of the fishery and suggested that a fishery in the Gulf of Mannar generates one off Sri Lanka and vice versa. Two hundred drift bottles were released on the Cheval Paar and fifteen out of thirty-five recovered were found on the

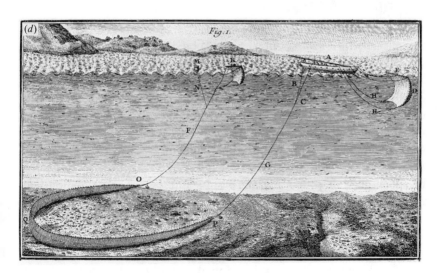

Figure 7. Seines: (*a*) shore seine hauled by men; (*b*) shore seine hauled by horses; (*c*) a seine in the open sea; (*d*) a bottom seine from which the Danish seine might have developed (Duhamel du Monceau 1769).

Figure 8. A simple trawl (Duhamel du Monceau 1769).

Indian coast near Tuticorin after fifteen days; the reverse drift was also established, during the south-west monsoon. Hornell showed that oysters spawned during the north-east monsoon and found that many banks swarmed with young settled oysters. They were exploited by *Balistes*, *Lethrinus*, *Serranus*, *Tetrodon*, rays and sharks; masses of fish gathered on three- to six-month old fishes. The adults are eaten by *Rhinoptera* and *Ginglymostoma*. The predation may have been enough to prevent a steady yield, accounting for the sporadic nature of the fishery. Hornell (1916*b*) quoted a letter by T. H. Huxley in 1864 which refers to the disappearance of young and old oysters and the recommendation that oysters be cultivated.

The chank industry (Hornell 1916b)

The Indian conch or chank was found in five places, Tinnevelly (or Tuticorin), Ramnad (north towards Madras), the coast of Karnataka, Travancore and Kathiawar (or Gujarat). Of these, the fishery in the Gulf of Mannar on the north-east coast of Ceylon was the most important. The sacred conch is a large gastropod which lives on sandy banks and eats terebellid polychaetes. The chank was cut into bangles, which were lacquered red and used as symbols of Hindi marriage in Bengal.

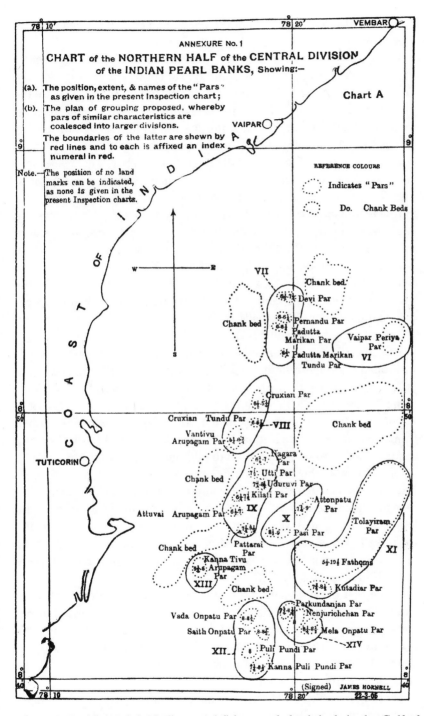

Figure 9. Positions of the Indian pearl fishery and chank beds in the Gulf of Mannar between India and Sri Lanka (Hornell 1922).

Necklaces also were made from the chank and there were many social uses for it.

The fishery for the chanks was prosecuted 1800 years ago under Pandyan rule at Korkai, 19 km south of Tuticorin. The Portuguese, the Dutch and the British were responsible for the fishery as they were for the pearl fishery. The fishery was sometimes difficult, because of headwinds, rough weather, turbidity, chilly water, morning calms, shoals of stinging jellyfish and the counter attraction of a sporadic pearl fishery.

The season opened in the middle or end of October. The divers were given an advance for ropes, but provided their own canoes and stores. The divers worked between early morning and early afternoon; they had to row their canoes out for 9–19 km. The crew comprised six divers and one thodai who tended the divers' ropes; there was no second rope as in the pearl fishery and the thodai hauled up the rope after the diver had gone down. The divers prospected from a drifting vessel and when catches were good the boat was anchored. They found the chanks by the tracks they made in the sand. The divers would get up to eight shells per dive and they might have made twenty-five descents each day, perhaps 100 chanks day^{-1} (Hornell 1914).

The chank beds were on fine sands with some mud and also on sands with dead madrepore branches in 16–20 m. Each shell weighed about 250 g; some shells were wormed, burrowed by *Clione*. On the Coromandel coast, up to 200000 chanks were taken each year. On the coast of Karnataka, the chanks were caught fortuitously by the net fishermen from catamarans. In the Kathiawar the chanks were picked off the shore at spring tides and some were sold to pilgrims.

Off Sri Lanka, the divers took live chanks and dug for subfossil ones. They worked in 6–10 m; it was so shallow that stones were not needed and they just filled their bags. The subfossil ground in mud was probed with long iron rods armed with hooks. About the same quantity of live and fossil shells was taken each year. Although the chanks were found on the north-east and north-west coasts of Sri Lanka, the best grounds were around Nayinativa Is. and Mannar Is.

In the first decade of the present century, the centre of the bangle-cutting industry was in Bengal, with Dhaka as the chief centre. Former bangle factories were in the Deccan, in Gujerat and Kathiawar; archaeological remains were found in Mysore, Bellary, Anantapur, Guddapah, Kurnool, Hyderabad, and Kathiawar. In recent history, most of the chanks came from Sri Lanka.

There were eleven trade varieties, handcut and lacquered. The bangles were used by Hindu Bengalis in Assam, Bahar and Orissa.

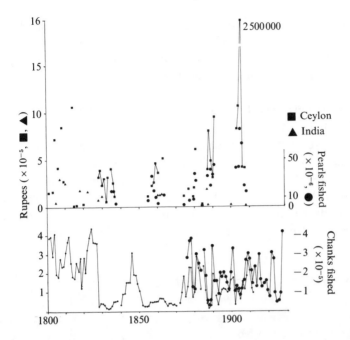

Figure 10. The yield in thousands of rupees from the pearl and chank fisheries in India and Sri Lanka and in numbers from Sri Lanka (Hornell 1922).

Inshore fisheries of France

The fisheries to be described in this section are those inshore; the major fisheries for cod, herring, tuna and sardines are described in other sections or chapters. Figure 11 shows three typical inshore vessels (*a*) fishing boats from Cayeux, (*b*) a Provençale boat in 'rough' weather, (*c*) a *gondole*, a somewhat larger vessel used for towing a trawl or *gangui*. Duhamel du Monceau (1769) describes the fisheries throughout the year by ports. From Dunkirk, whiting were caught in December and January. Between 1 February and 15 May larger boats sailed *c*. 200 km northward (to the region south of the Dogger) for cod, haddock and rays. Up to the end of July only rays were taken. From the end of September onward, herring (and some mackerel) were caught from the Channel and salted in Dunkirk. A most remarkable observation is that halibut were found on the coast of Flanders, but most commonly on the Dogger Bank at weights of 11–45 kg (although fish as large as 180–317 kg had been taken).

From Calais, rays, whiting, lemon sole, flounders and shrimp were caught all the year round. Brill were abundant between April and the end

of summer. Mackerel were brought in between 12 May and 15 July.
Weevers were taken by dredges inshore in hot weather, but in deeper
water on cool days. Herring were caught between 12 October and the end
of November. From Boulogne, demersal fish were caught all year round
by dredges and trammels. Mackerel were taken between May and July in
500 m drift nets. The herring fishery lasted between 10 October and
10 December.

In High Normandy (between Le Tréport and the Seine), demersal fish
were caught by long lines, dredges, small beach seines, fixed gill nets and
trammels 500 m in length. Monkfish were caught in June and July, but
turbot were rare, as indeed they are today. Herring and mackerel were
fished by drift nets. In Lower Normandy, west of the Seine, trammels
were used in April for plaice, sole, brill, small turbot, dogfish, lobsters

Figure 11. Some early French fishing boats: (*a*) fishing boats off Cayeux; (*b*) a
Provençale boat in 'rough' weather; (*c*) a gondole (Duhamel du Monceau 1769).

and octopus. Between October and April, whiting, plaice and dogfish were caught on lines, lesser sand eels in small seines. Spent herring were taken between mid October and St Catherine's Day (29 April).

In the Baie de la Hougue, sole, plaice and brill were caught by the trammels. Between 15 July and 15 October lemon soles, whiting, congers, saithe, red mullet and gurnards were prominent in the catches. Shrimp were caught between June and September and lobsters were taken between April and October. At St Malo in Brittany, saithe appeared amongst the demersal catches; ormers were also recorded. At St Brieuc, not far away, plaice, sole, ray, bass and mackerel were caught in May, June and July; in July, August and September, catches comprised big rays, gurnards, soles, mullets, bass, saithe, whiting, plaice, turbot, congers, and spotted dogfish.

In Southern Brittany, sardines were caught with 400–480 m drift nets at Concarneau in June, Douarnenez in August and Brest in September. Cod and mackerel roe were used to attract the fish in daylight (see below). Between April and May, eels were captured with the aid of lights. Congers were caught between Audierne and the Ile des Saints; hake were taken on hooks and saithe were abundant. In La Rochelle, sardines were fished between April and October and bass between April and June.

In Collioure on the Côte Roussillon, bluefin tuna were trapped in *madragues* and were sighted from towers. Sardines were fished by drift nets up till an hour after sunset and again at dawn. Duhamel du Monceau (1847) listed species by ports but his description of seasonal fisheries in the Mediterranean is sparse, perhaps because there is less seasonality there.

The Breton sardine fishery (Cadoret *et al.* 1978)

The *sardine de rogue* fishery is worked in daylight with cod roe (*la rogue*) scattered on the sea. It is probably about 300 years old, and ancient texts speak of a fishery between Camaret and Sables d'Olonne; between 1640 and 1650, 40 000 barrels (each of 3000–6000 fish) were landed each year. The oil press was developed in the eighteenth century and the products were distributed all over France and to many parts of the Mediterranean. The first cannery was opened in Nantes in 1822 and many others were built on the southern coast of Brittany in the 1830s and 1840s. The demand from the canneries stabilized the market, and the oil presses and the fresh fish wagons went out of business. The fishery was steadily productive until 1902–8, when it collapsed; after the First World the decline continued.

The fishery started on 15 June (if earlier the market would have been depressed). At Notre Dame de Larmor was celebrated *la messe des sardines*; on 24 June, the fleet was blessed at sea. At Douarnenez and Tréboul, fishermen's wives visited the fountain of Saint Per at Paoul after dinner on a pilgrimage. They burned candles and prayed for good fishing. If the fishery was poor, they took the saint's figure, whipped it and laid it on the grass face down.

The first boat to find sardines was called *le découvreur* and the first sardine, *le bouquet*, was eaten in wine or cider (in 1864). The fishery lasted until St Martin (11 November); in mid fishery at St Michael, a local pardon was celebrated. In the first half of the nineteenth century, nets were: 45–50 m long and 270 meshes deep at Belle Isle and Port Louis; 32 m long and 240–50 meshes deep in Concarneau; and 26–27 m long by

240 meshes deep in Douarnenez. Towards the end of the century there were three standard nets, 55 m in Arcachon, 35 m in the Vendée and 45 m in Brittany; each boat was equipped with two sets of fifteen nets. The patron chose the nets he needed and weighed them down with a 2 kg stone to a depth of 10 m. He searched for signs in the water, its colour or oil on the surface, predators such as porpoises, dolphins and albacores and competitors such as fulmars, puffins, gannets and guillemots; he tended to fish at regular positions. He gave the sign to *jetez dehors*, 'les casquettes en bas pour bénir le coin de pêche'; this custom tended to die after the First World War.

The boy made the *rogue*, cod roe, in peanut flour (formerly fine sand). In former times, each crewman dipped his finger in a little bottle of holy water and the patron scattered a few drops over the gear. The *rogue* sinks, leaving an oily wake for some hundreds of yards; it was scattered under the sign of the cross. If there was no fish, the net was hauled, mast raised and sail set to search again; this operation might have been repeated three or four times a day. The crew watched for fish in the water, sometimes with the boy on the end of the mast. The fish rise to 7 or 8 m and, when the fish were caught, the patron signalled to the rest of the fleet.

In 1860 the boats could be recognized by their shape and rig but, in an extraordinary expansion, boats were built at Le Croisic, Palais, Locmalo, Quimper, Douarnenez and Concarneau, all of the same pattern. The number of boats built in Douarnenez were:

1882: 67	1897: 141	1901: 148
1883: 79	1898: 145	1902: 127
1884: 69	1888: 169	1903: 102
1896: 122	1900: 176	

At the turn of the century, there were 500 boats at Concarneau and 800 at Douarnenez and, in 1911, 3200 along the whole coast.

The Breton tunny fishery (Cadoret *et al.* 1978)

In 1727 there were twenty decked boats fishing for tuna from the Ile d'Yeu, and during the later eighteenth century the albacore fishery followed that for sardines. Between 1840 and 1860, the fishery was based on Les Sables d'Olonne and the Ile de Groix; indeed the population on the island increased with the fishery. In 1817, there were twelve small boats working from the Ile d'Yeu; in 1893, there were 500 between Camaret and La Rochelle and 874 in 1934. The essential development was the introduction, in the 1860s, of the *dundée*, an adaptation of the smack with a dandy rig, based on Yarmouth and Lowestoft. It was 15–25 m in length; during the winter it trawled with beam trawls off the southern coast of Brittany. A form of beam trawl had been used for centuries on that coast.

In the spring the trawl gear was landed, the ballast was unshipped, the hold washed, brushed and fumigated with burnt sulphur to kill small crustaceans; then the ballast was put back. The winter sails were replaced by those of summer. The boat was repainted and the tangons (the trolling rods) with spars and halyards were shipped. In mid June, the *thoniers* sailed after the feasts of departure. Between 1896 and 1909 they worked up to 480 km from port, and made voyages of fifteen to eighteen days duration. During the period from June to September the fishery shifted from south to north. The shelf edge was the best place, where there were 'heavy swells', perhaps associated with internal waves. For navigation they used three main shipping lines across the Bay of Biscay.

The crew (skipper, four men and cabin boy) scattered maize and horse-hair at the surface to attract the albacore. The *dundée* with its sails full with wind travelled at 9–11 km h^{-1}. Little bells on each line signalled the fish on the hooks. The fish fought long and the fishermen urinated on their hands to avoid cracks and wounds. The shoals of albacore were crossed in a few hours and the skipper turned in a broad circle to avoid tangling the lines from the tangons. If the wind strengthened, the number of lines was reduced to two. When landed aboard the fish were hung in pairs from a horizontal centre pole from the foremast. They were wetted with sea water and sometimes protected from the sun by tarpaulins. The fish were sometimes taken ashore by special carriers. Freezing fish at sea was introduced in 1933. Ashore the first cannery for tuna was built at Groix, but later Concarneau became the most important centre.

The fishery reached its peak in the 1920s and 1930s. After the Second World War it continued though in decline, but the sailing vessels were still seen at sea as late as 1960. In the summer of 1934 I saw the Concarneau fleet standing in from sea becalmed in the evening when one boat slipped into port on its new engine. The post-war history was one of the development of motor trawlers, which could also troll for tuna.

The tuna fishery in the Mediterranean and its approaches (Parona 1919)

There are two fisheries in May and June and between July and early autumn. The first exploits the spawning migration into the Mediterranean and the second that of spent and young fish. The fish were caught by trolling in the Straits of Messina, by small traps (*tonnarelle*) with *vedette* or sighting towers in Dalmatia, on the Roussillon coast and at Sète in France and in Algeria. There were special very long trap nets for albacore. But the main trap net for bluefin was the *tonnara* (Italian), *almandrabai* (Spanish), *armaçao* (Portuguese), or *madrague* (French). There were various varieties, *tonnare di corsa* (spawning migration), *tonnare di ritorno* (autumn migration), *di sopravento* (windward) and *di sottovento* (leeward). Those in Sardinia and Sicily are submarine castles, 200–400 m in length, 40 m broad and 20–50 m deep.

Figure 12 shows a typical *tonnàra*. In this case, the fish migrate along the coast from the east and are turned seaward by the guiding wall (which may be 120 to 3500 m in length); the tuna swim into the entrance and are kept in the eastern compartments. When needed they are brought into the western compartments, ending in the 'death room'.

The crew of a *tonnara* is directed by the *rais*, who has the sureness of eye and promptness of manoeuvre to supervise the building; he also needs exact knowledge of the sea bed and of the influence of wind and current. He sets the nets properly and estimates the quantity of fish entering the trap. He has one or more assistants. The crew comprises a number of specialists, anchormen, boatmen, capstan boatmen, etc. The *tonnara* off Trapani was manned by a crew of ninety-three. Ashore at Trapani, the crew are supported by porters, stevedores, cooks, boxmakers, coopers, etc.

On 3 May, the crews sail at dawn to spend several days to set up the nets. When the fish enter the trap they are brought to the surface with white cuttlefish bones suspended in the water. When enough fish are in the eastern compartment, the door is opened and the fish are herded towards the 'death room' with white palm leaves. The *rais* orders the fishery to start and he stations himself in a little boat in the middle of the 'death room'. He gives the signal to fish 'in the name of the Lord' (as the East Anglian driftermen said) and the net is hauled in to the rhythmic chant of the *aimola* (or shanty). As the net brings the fish to the surface they are killed by lances in the *mattanza*, or death room.

Figure 13 shows the distribution of *tonnare* in Italy, Sicily, Spain and Portugal, and Turkey. The distribution is of considerable interest, particularly on the Algarve and the Atlantic coast of Spain, the tourists' Costa Blanca, the coast about Marseilles, the Hellespont, Elba, Porto San Stefano, the coasts of Sicily and the extraordinary sector from Rijeka (Fiume) southwards on the Karstish coast of northern Yugoslavia. Between 1890 and 1914, there were in Italy 37–54 *tonnare*, 2300 to 3700 crewmen employed and 16237 to 81597 quintals caught (3000–19000 tonnes net weight). The fishery was relatively small, but was always

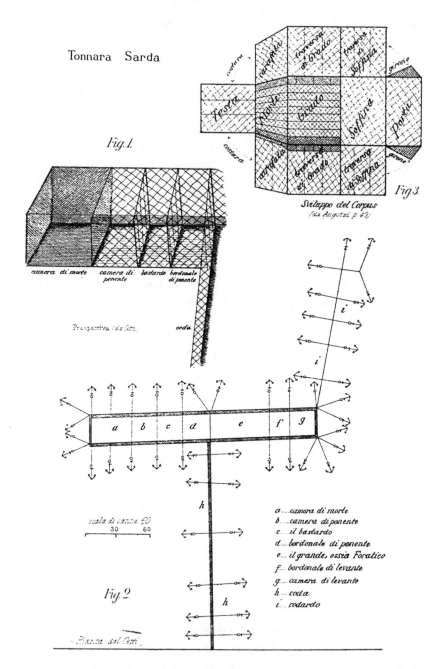

Figure 12. A typical *tonnara* (Parona 1919).

valuable. A large number of *tonnare* were abandoned by 1919; the decline was attributed to a number of causes, more shipping, malaria, the capture of young fish or more generally, overfishing.

The preindustrial fisheries of the United States

Introduction

In 1877 Brown Goode put forward a scheme for recording statistics of the fisheries and for describing them. In 1879 he was asked to execute it as part of the Tenth Census. The results were published in his *History and methods of the fisheries* (1887); four other volumes recorded other aspects (Brown Goode 1887). The five volumes provide one of the fullest descriptions of preindustrial fisheries in existence. Of course, industrialization had started long before in the United States and by 1877 the railroads linked many cities. The market for fish had expanded particularly after the Civil War and there were canneries for Pacific salmon and the Maine herring. But the methods of capture were preindustrial and, however developed and lovely the vessels, they were driven by sail.

The temperate fisheries of New England

The major fisheries for halibut, cod, haddock, mackerel and herring were based on the ports of New England, particularly Gloucester. Not far away were the new centres of industry, accessible by railroad.

The fresh halibut fishery (Brown Goode and Collins 1887a)

In the early nineteenth century, halibut were very abundant in Massachusetts Bay. Before 1825 it had been considered a pest but demand in Boston generated a fishery with handlines on George's Bank from 1830 onwards. Between 1828 and 1848, the fishermen worked on George's Bank from Gloucester (Figure 14). By 1848, sixty-five vessels were landing as much as 13–25 tonnes each in the season. Between 1850 and 1852, this fishery ceased to be profitable and the fishermen shifted to Seal Island Bank, Brown's Bank and Sable Island Bank (Figure 15). In 1850, the year the 'halibut gave out', great pieces of halibut were found in the throats of cod.

The further development of the fishery was stimulated by the construction of the Providence and Stonington railroad. In 1849, trawls (which I shall call long lines) were introduced by Atwood from Provincetown and by Sinclair from Scotland. In 1846, the first ice house had appeared on a schooner. When the windlass was patented in 1849, a five hour haul with handspikes was reduced to half an hour. In 1850, salting was introduced in a small way.

In 1879–80, the Gloucester fleet comprised two-masted schooners with heavy anchors and anchor cables of 750–850 m in length. they carried much stone ballast, salt and ice. Each vessel carried six dories; each used six long lines with 350–380 hooks on each. They were baited with herring,

(a)

ITALY

SICILY

mackerel, menhaden, cod, haddock and hake. the lines were set under sail in daytime and were hauled on the windlass fifteen to eighteen hours later. The fish aboard were handled by two gangs of three, cutter, blooder and icer; the fish were gutted, cut and thrown on to ice in the hold.

The Gloucester fleet worked in deep water (except on the west coast of Newfoundland) between Georges Bank and southern Labrador. From New York, the fleet worked in 50–200 m off Nantucket, on George's Bank, Brown's Bank, Seal Island Ground and Sable Island Bank. In

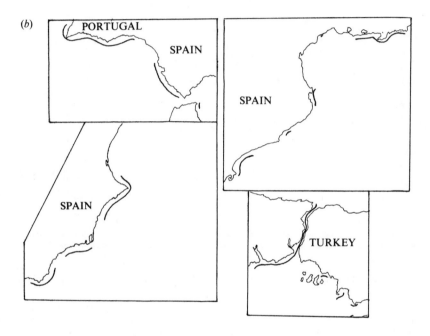

Figure 13. Distribution of *tonnare* in the Mediterranean shown by a full line off the coast. (*a*) Italy and Sicily; (*b*) Spain, Portugal and Turkey (after Parona 1919).

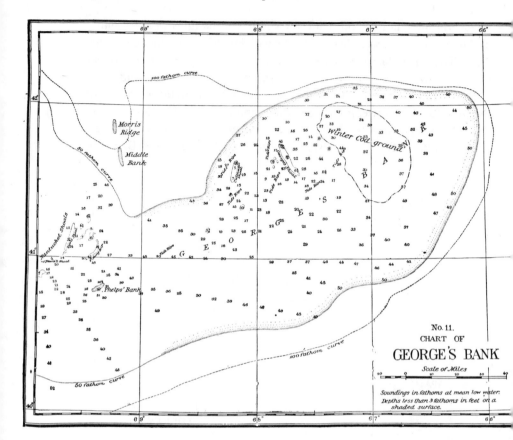

Figure 14. Chart of George's Bank (Collins and Rathbun 1887).

1878, about 4852 tonnes of fresh (or iced) halibut were landed in Gloucester. During the 1870s a small fishery for smoked halibut was sustained in the Davis Strait.

The cod fisheries (Brown Goode and Collins 1887b, c, d, e)

There were two fisheries for cod from New England ports, that on George's Bank and that on the Grand Bank, Gulf of St Lawrence and off Labrador.

Cod was discovered on George's Bank (Figure 14) in June 1821 but the fishery from Gloucester really started after the decline of the halibut in the 1850s. The vessels were schooners with a 450 m anchor cable and they carried one dory on the square stern; they carried 7–10 tonnes of ice and

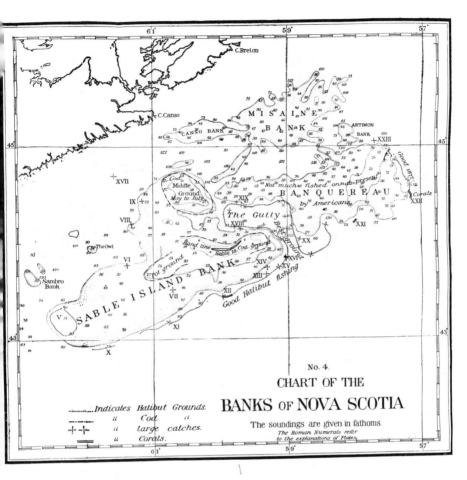

Figure 15. Chart of the banks off Nova Scotia (Collins and Rathbun 1887).

some salt. The crew comprised eight to twelve men and the fishermen worked with handlines from the deck (Figure 16). Each line was 300 m long and hooks were fastened by a special George's gear. The hooks were baited with frozen herring in winter and with fresh herring, mackerel, alewives and menhaden in summer. The lines were fished mainly in daytime and sounded to the bottom.

In 1855–8 hand-lining for cod from dories was started. The schooners carried eight to twenty dories nested upside down on deck to the region of the Grand Bank. No ballast was carried and the hold was filled with salt (on sailing). The vessels were large and indeed were often used in the carrying trade in winter. The anchors were lighter than those used in the

Figure 16. The George's Bank cod fishery (Brown Goode and Collins 1887*e*).

Figure 17. The dories and schooners on the Bank trawl (or long) line fishery
(Brown Goode and Collins 1887*d*).

Figure 18. Dory crew of cod fishermen catching birds for bait (Brown Goode and Collins 1887*b*).

halibut fishery. Handlines of 100–150 m were used with leads of 1.4–2.3 kg, with single snoods because the dories often fished close to each other in less than 90 m (Figure 17).

The lines were baited with salted clams, fresh capelin, fresh squid and sometimes even birds (Figure 18). On the Bank, the schooner laid on a riding sail and the dories started fishing at sunrise; they came back at midday for dinner and then went out again till dusk. The schooners left Gloucester in April, May or June and worked the Grand Bank, but they also fished the shoal waters of the Banquereau and Sable Island Bank (Figure 15).

In 1880 there were 200 American schooners working long lines on the Grand Bank from dories. The vessels left Gloucester in February and March for Sable Island Bank, but for the Grand Bank they sailed in April, May and June. The four to six dories used 150–200 barrels (18.75–25 tonnes) of bait in the season. In the early part of the season they worked the southern part of the Bank, but in July and August they spread their effort more widely from Brown's Bank to St Pierre (Figures 14 and 15).

In 1879–81, an average of 10 491 tonnes of cod were landed from the George's Bank fishery. For the same three years an average of 7862 tonnes was landed at Gloucester.

Figure 19. The haddock fishery on George's Bank; the lines are set at right angles to the course of the vessel (Brown Goode and Collins 1887*f*).

The haddock fishery (Brown Goode and Collins 1887f)

In 1850 vast quantities of haddock were found in Massachusetts Bay and the fishery dated from this period. Between October and April it extended from Sandy Hook to Sable Island in 50–180 m. But the Gloucester fleet worked on George's Bank, Brown's Bank and La Have in 50–80 m. The schooners resembled the halibut vessels but they rarely anchored. They carried six dories which were dropped off under sail. The long lines were 600 m in length, each dory carried six to eight tubs and each tub carried 500 hooks. The lines were set once a day and it took up to four hours to haul them (Figure 19). Haddock were less liable to drop off a slack line and the fishermen had to be careful how they managed them. The schooners carried as much as 5 or 6 tonnes of ice and on capture the fish were gutted and washed before being laid on ice. The voyages lasted two or three days and because the fish quickly lost quality, the schooners had to rush for port; vessels were sometimes lost. In 1880, catches probably amounted to 7763–8734 tonnes.

The mackerel fishery (Browne Goode and Collins 1887g)

The early methods of catching mackerel were by beach seines along the coasts of New England, or by trolling with long poles. Jigging with

Figure 20. The mackerel fishery (Brown Goode and Collins 1887*g*).

weighted hooks was introduced in 1816 and between 1820 and 1870 the hook fishery was a considerable one, predominant until 1860; after that date the purse seine displaced it. The hook fishery extended from the Chesapeake to the Gulf of St Lawrence. Between March and the end of May, the vessels worked as far north as the south shoals of Nantucket; in this spring fishery, the schooners brought their catches into New York two or three times a week. As they moved away from the Delaware towards the north, the schooners fished to within a kilometre or so of the beach. In the summer and fall they operated in the Gulf of Maine.

In the first decades of the fishery, the vessels were the old-fashioned square-stern schooners, but swifter boats were introduced in 1855. The decks were clear and long timbers were mounted on the rail to carry the cleats for the lines and the bait boxes. The fish were split, cleaned, washed, cured in very fine salt and packed in barrels in the hold. The hooks were baited with pork rind, but menhaden and mackerel were also used for chumming, i.e. scattered on the surface to attract the fish. At the masthead stood a lookout searching for 'a spurt of mackerel'. Figure 20 shows the men catching the mackerel and Figure 21 illustrates the catches in Massachusetts between 1804 and 1881; until 1850 or 1860, the catches were taken by the hook fishery.

The first purse seine was made in 1826; it was 130 m long and 4–6 m deep. But the mackerel seines were not introduced until 1853 or 1854; the earliest working seine was adapted from a whaler, but it had a sharpened bow; it was stiff in the water and had a stern platform. It was towed 40–100 m astern. Most of the larger vessels carried two seine boats and two

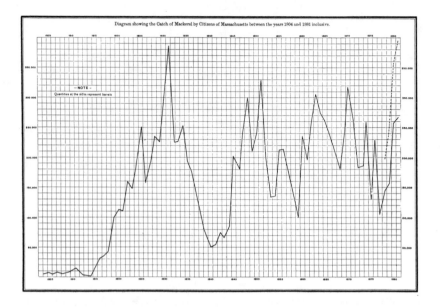

Figure 21. Catches of mackerel off Massachusetts (Clark 1887*c*). The broken line indicates the total quantities of mackerel taken, whether sold in the markets in a fresh condition or salted.

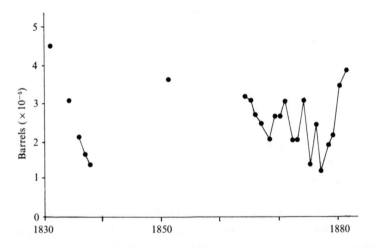

Figure 22. Catches of mackerel taken off the eastern seaboard of the United States between 1831 and 1882 (Brown Goode and Collins 1887*g*).

seines, a large one (450 m in length, 40–50 m deep) and a smaller one (300–500 m long and 20–24 m deep) for use in shallow water (see Figure 51, p. 121) (Brown Goode *et al.* 1884).

The schooners, like the hook boats, started to work in spring between the Chesapeake and the Delaware and fished their way north to the Gulf of Maine. The skippers used salted menhaden as a chumming bait before shooting the seine. Figure 22 shows that the purse seine catches predominated after 1865 or 1870. The American purse seine was tried in Norway but it failed there. In 1880, 57 156 tonnes of mackerel were landed in this fishery; in the same year the British Provincials (or Canadians) landed 31 873 tonnes. There were 468 vessels in 1880, 338 purse seiners, 81 liners, 44 gill netters and 5 liner/purse seiners. Most of the catch was pickled in salt and barrelled, but part was also canned.

The herring fisheries (Earll 1887a)

The herring were fished between Cape Cod and Eastport (on the American border with Canada); the fish appeared off Cape Cod in April and a few weeks later off the coast of Maine. Great quantities were caught on George's Bank. The fish were caught by 'torching' (with dip nets at night under the light of torches (Figure 23)), by brush weirs (traps on the shore line (Figure 24)), gill nets and purse seiners. The brush weirs had

Figure 23. Torching, a method of catching small herring at night (Earll 1887*a*).

Figure 24. The brush weirs of the sardine (or young herring) fishery along the
shores of Maine (Earll 1887*a*).

long leaders and funnel-shaped openings, were sited on extreme points of
land or in a channel between islands; the fish were taken by beach seine
within the trap. The gill nets were 30–40 m long, 4–6 m deep and the fleet
comprised eight to twelve nets, manufactured of cotton. They were
suspended in the water between floats and sinkers.

There was a variety of fisheries for frozen herring, pickled herring,
smoked herring and 'sardines' (immature herring). That for frozen
herring was prosecuted by the largest and most able schooners (because
of gales and ice in the depth of winter). In 1854, herring frozen on deck
were brought back from Newfoundland to Gloucester. The fishery took
place in Fortune Bay on the southern coast of Newfoundland. The
schooners reached the Bay in mid December and left in mid January. In
the early days, gill nets were used, but later beach seines were employed.
The hold was built up into an ice house where the fish were held. Many
vessels were lost in the entrance to Fortune Bay and others were lost in ice
or under the 'herring press of sail'. By 1880, about 5822 tonnes of frozen
herring were landed in Gloucester for bait in other fisheries.

The fishery for pickled herring took place for three or four months in summer in the Gulf of St Lawrence off the Magdalen Is., off Anticosti, off West Newfoundland and off southern Labrador. In the early stages of the fishery (1822), small schooners worked with gill nets and beach seines. The fish were salted in bulk and not packed into barrels until the vessels returned to Gloucester. After 1865, larger schooners took part and they carried up to 400 hogshead of salt. By 1872, the bulk of the catch was taken by purse seine. In 1839, perhaps 100 000 barrels were packed, but by 1876 only 77 443 barrels were produced.

Along the coast of Maine, herring were smoked. Before 1828 the fish were taken by torching, and in that year the first brush weirs were built in the Quoddy estuary and in 1877 or thereabouts gill nets were introduced. Smoke houses were built all along the coast between Eastport and Portland. The fish were scaled, washed, salted, drained and hung in the smoke houses for two to six weeks, depending on size. In 1880, 307 300 boxes of 'hard' herring and 51 700 boxes of bloaters left the coast.

The brush weirs (Figure 24), which came from New Brunswick, depended largely on immature herring. The fishery lasted from September till the end of the year. In 1849, there were sixty-five weirs on the American shore and fifty-five on the 'English islands'. By 1879, there were ninety-nine in New Brunswick and sixty-six on the American shore. A canning industry was started in 1875 and by 1880, 6 284 300 cans were sold.

The fisheries of New England

The five fisheries for halibut, cod, haddock, mackerel and herring have been described separately for they are distinct entities. But to some degree the fishermen and their vessels overlapped, switching from one fishery to another. Much of the herring fishery was restricted to the ports of northern Maine, but, like the other fisheries, some herring fishermen were based in ports such as Gloucester. From the five fisheries, 35 779 tonnes were landed there. As industrialization in the United States proceeded, it drew the fish from the sea to feed the rising towns. The price is given in Figure 25, which shows the number of fishermen and the number of vessels lost each year from the port of Gloucester in Massachusetts.

The swordfish fishery (a subtropical fishery from New England Brown Goode 1887b)

Swordfish were taken between New York and the coast of Maine; they appear in June off Sandy Hook and Block Is. Sound and by July they have

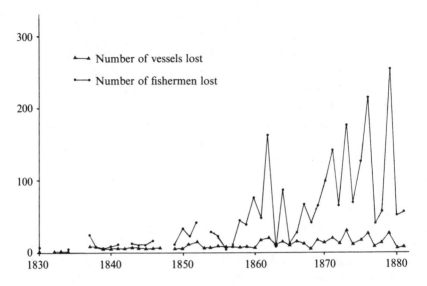

Figure 25. The number of fishermen lost (●) and the number of vessels lost (▲) from Gloucester, 1830–61 (after Brown Goode *et al.* 1884).

spread between Montauk Point and the coast of Maine. By August and September they are between George's Bank and Sable Is. Presumably the swordfish are carried north in the Gulf Stream and individuals are spun off towards the offshore banks.

The gear was a harpoon with a detachable head, a hickory pole 4.6 or 4.9 m long. The harpoon line was 100–300 m in length. The skipper of the sloop or small schooner stood in an iron 'frame' or 'pulpit' above the bowsprit (Figure 26). Each harpooner carried a lance. The fish could be seen at the surface in calm or moderate weather; indeed from the mast-head the dorsal fin is visible at a range of 3–5 km in calm weather. The schooner could approach the fish more closely than a small boat. The harpoon was released at the very close range of 180–300 cm. Then the fish was brought in on the line to the small boat. The operation was exciting or dangerous, for the swordfish could attack vessels; such a fishery can be seen today off Scilla, just north of the strait of Messina.

The fishery started in the late 1830s and the number of vessels ranged from twenty to fifty. In 1861, there were thirty schooners working from New Bedford. A certain number of mackerel purse seiners started to carry 'irons'. The *Northern Eagle* made eight trips in one year and caught 163 fish, or 20 tonnes. Figure 27 shows the catches of swordfish pickled in numbers in Massachusetts between 1840 and 1877; the total catch from the coast was, of course, higher, perhaps about 2000 in the 1870s.

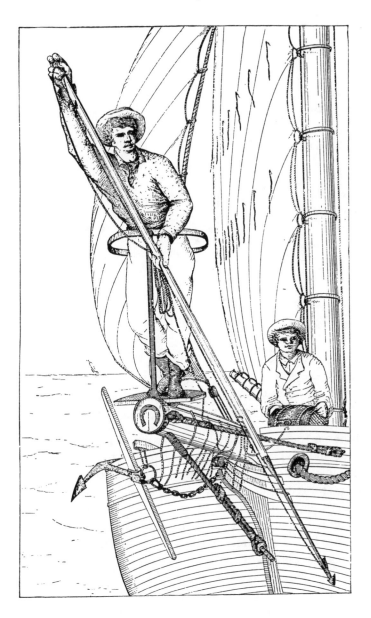

Figure 26. The swordfisherman with his lance (Brown Goode 1887*b*).

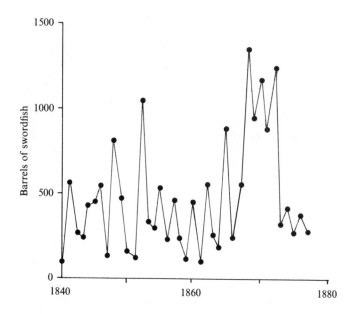

Figure 27. Catches of swordfish from 1840 to 1887 (after Brown Goode 1887*b*).

Some fisheries between Sandy Hook and Mexico
(Collins 1887; Earll 1887b; Stearns 1887)

The fisheries south of Sandy Hook were not as productive as those off New England, and the species of course tended to be more subtropical. For example, schooners fished in the Gulf of Mexico from Pensacola (in Alabama) for red snapper. They worked in gullies in depths of 20–80 m; from Pensacola, in summer they fished over 30 km and 120 km in winter. The fishermen used lines 120 m long, with two hooks and a sinker; the hooks were baited with fresh or salted bluefish and ladyfish. In 1885–6, thirty-eight vessels practised this trade.

In all bays between Sandy Hook and Mexico, mullet were caught by beach seines (658 in North Carolina), gill nets (966 in North Carolina) and cast nets (285 in East Florida), a diffuse spread of continuous fisheries. Pamlico Sound and the estuaries near Charleston were the main centres from which 1286 tonnes of fresh mullet and 2477 tonnes salted were landed (Earll 1887*c*).

Spanish mackerel were fished by trolling gear, gill nets and pound nets (a fixed engine or trap). The trollers worked in large open boats between Long Is. and New Jersey. Gill nets about 200 m in length were used between Sandy Hook and the Chesapeake. The pound nets were set along the shore between Connecticut and the Delaware; off the Chesapeake the fish were taken between May and September and off Long Is. and New Jersey in August and September. On the shores of Delaware Bay in shoal waters 'sharpie' skiffs with large, small and wade beach seines were sailed to catch squeteague, spot, rock fish and perch.

There is a contrast between such fisheries and those in the north off New England; they were small, diffuse, spread from port to port, and supplied an array of rural communities. In the north, fish were drawn off the Banks into the burgeoning cities.

Two fisheries on the west coast

The Alaskan cod fishery (Bean 1887)

The first cod was brought into San Francisco from the island of Sakhalin in the Sea of Okhotsk in 1863. Large schooners sailed the 6400 km by way of the Aleutian Is. In 1868, cod were found near the Shumagin Is. at the seaward end of the Alaskan peninsula. Subsequently this fishery was exploited by smaller schooners. Hand lines and long lines were used from dories and they were baited with clams, herring, sandeels and squid. The Shumagin fleet extended its range eastward to Kodiak Is. and westward to the Aleut passes. The average catch landed in San Francisco between 1865 and 1880 amounted to about 970 tonnes of dressed fish; the catch was salted in the hold, 'bundled', boxed boneless. Only about ten or twenty vessels were employed. The Shumagin fleet caught on average 58000 cod each year with each vessel, whereas that from Okhotsk took 166000 cod per vessel annually. In 1878 the catches by the two fleets were about the same, but by 1880 the Okhotsk catch was three times that of the Shumagin.

This transpacific fishery supplied a market in San Francisco. In 1865, seven vessels landed 570 tonnes. Between 1876 and 1880, an average of nearly 1456 tonnes was brought in each year by about eleven vessels. It was a small fleet to go so far.

The Pacific salmon fishery (Jordan and Gilbert 1887)

All given species of Pacific salmon, the sockeye, the chum, the king, and the coho, were caught in the Sacramento, Smith, Eel, Rogue and Columbia rivers. The Sacramento is one of the two major rivers which empty into San Francisco Bay; the Smith and Eel are small rivers in northern California, the Rogue in Oregon and the Columbia flows to the sea between Oregon and Washington State. Gill nets, dip nets, traps and beach seines were used. The gill nets were 400–600 m long and 12–18 m deep. They were drifted downstream and caught the fish as they swam upstream to spawn. A proportion of the catch was salted, but most was canned. The first canneries were built on the Lower Columbia and, in 1880, half a million cases (or cans) were produced. The king salmon was the chief species taken in the Sacramento; in the Eel, Smith and Rogue, king and coho were the main species; all five species occurred in the Columbia. The catches in these rivers, in Puget Sound and in Alaska in 1880 were:

	tonnes
Sacramento river	2331
Smith, Eel and Rogue rivers	699
Columbia river	15649
Puget Sound	87
Alaska	295
Total	19061

Across the border in British Columbia 42155 cases were packed in the Fraser and 19694 in the Skeena, 61849 in all or 1304 tonnes. Thus, at the beginning, the Canadian industry was dwarfed by the American.

Crustaceans, molluscs and turtles

Crustaceans (Rathbun 1887)

Blue crabs were taken by scoop net, dip net or hoop net as soft crabs in the Chesapeake and the Delaware. A crabber's trotline of 80–400 m in length was baited for the hard crabs. Horseshoe crabs were gathered on the coasts of New Jersey and Delaware. In 1880, 6480 tonnes were landed.

Lobsters were taken chiefly in Maine and Massachusetts, although smaller quantities were landed as far south as New Jersey. A variety of small boats was employed up and down the coast and the lobsters were caught in funnel traps and pots (made of hoops, laths or net), baited with herring, sculpins and flounders. The animals live in relatively deep water in winter and migrate into depths of 12–20 m in summer where the fishermen find them. In 1880, 8154 tonnes were landed. By that date, the mean

sizes of lobsters had already been reduced to near the minimum prescribed by law.

Molluscs (Ingersoll 1887)

Oysters are found from the Gulf of Mexico to Massachusetts Bay. Since
the time of the early settlements many grounds had been swept clear; but
they were repopulated by laying new seed. On the shoals of the James
River in Virginia, men went 'seed tonging' in small boats, as many as a
hundred in 200 ha; the boats were dugout canoes, pointed at both ends.
The tongs were 30, 81 and 91 cm in breadth. Such seed oysters were
carried away to northern grounds. On these grounds, after the animals
had grown, they were fished with dredges, drag rakes and tongs.

Increased demand after the Civil War created avid markets in New
York and Philadelphia. In 1880, 782 122 668 l were landed mainly from
Maryland and Virginia; that is, the Chesapeake. There were 38 249 fishermen and 14 556 'shoremen'. The work was very hard and 'very few oyster-
men reach old age'.

Scallops are found between Cape Cod and Long Island Sound. They
were caught by dredges from small boats; the dredges were adapted to
work on soft and hard grounds. In 1880, 272 781 l of meat were produced.

Soft clams were dug with stout forks from the sand when the tide was
out. Between New Hampshire and New Jersey, 37 518 146 l were landed
in 1880. Quahogs were collected with clam rakes, straight rakes, curved
drag rakes, dredges and oyster tongs. In 1880, 38 320 940 l were landed
from the coast between New England and the Chesapeake. In the same
region, 21 142 860 l of mussels were landed. On the coast of California,
27 401 147 l of abalone meat and 1740 tonnes of shells were landed.

Turtles (True 1887)

Green, loggerhead and hawksbill turtles are found from Delaware to
Texas. In North Carolina they were attacked by spears until fishermen
learnt to dive overboard and steer them to capture. Further south, greens
were caught in gill nets, bigger ones southerly, 9–11 kg to more than 45 kg
each off Cedar Keys, but only 160 tonnes in all.

The preindustrial fisheries

The fisheries in 1880 in the United States supplied a long coastline of rural
communities and a rising network of east coast cities. The methods of cap-

ture remained those of earlier times, but they were supported in New England by thorough-going development on traditional lines. In later decades, the methods of capture themselves would be industrialized and the nature of the outputs would be altered. The quantities caught would be very much greater, particularly from the New England fisheries. The preindustrial fisheries did not land the enormous catches to which we are accustomed, and large numbers of fishermen were employed to make relatively small catches. The losses of men and of vessels were high.

3

The cod fishery off Newfoundland
1502–1938

The explorers who discovered Newfoundland, Cortereal and the Cabot brothers, reported the abundance of cod. From these reports the Portuguese and French fisheries started, to be followed by the English and later the American and Newfoundland fisheries. Through the centuries there is a history of conflict mainly between the British and French but also between the British and the Americans. Such conflicts were resolved far from Newfoundland, in European wars and in the American War of Independence. The histories written of the fisheries record the part played by them in the context of the greater conflicts. My view is much more limited, to describe the method of fishing and the quantities of cod caught. It will be shown that the quantities caught and exported to Mediterranean countries and the West Indies were indeed enough to be a bone of contention between nations. Figure 28(*b*) shows the grounds off Nova Scotia and Figure 28(*a*) is a map of the Newfoundland fishery (Innis 1940).

The French fishery for cod on the Grand Bank

Introduction

The French have fished for cod on the banks between Nova Scotia and Labrador since the sixteenth century. The fishery was large before the wars of the eighteenth century and it supplied home consumption and export. During the eighteenth century, there was a series of conflicts, the War of Spanish Succession, the Seven Years' War and the American War of Independence, each of which had distinct effects upon the French fishery. The information is taken mainly from de la Morandière (1962–6).

The nature of the fisheries

There were two fisheries, for salted cod and for dried cod. The cod for salting was taken by vessels working on the open banks and was returned

(*a*)

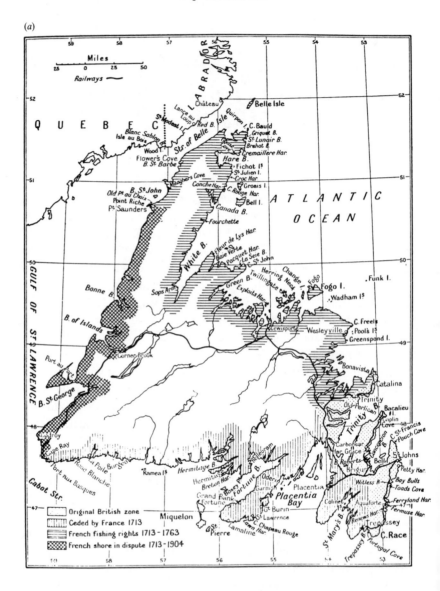

Figure 28. (*a*) Chart of Newfoundland, showing the areas of French and British influence; (*b*) chart of the fishing grounds (Innis 1940) (1 mile ≃ 1.6 km).

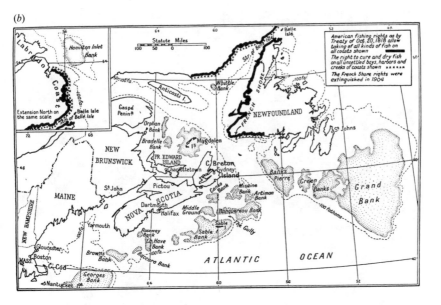

to home ports for consumption in France. The cod for drying was caught from small boats in a coastal fishery. The boats were carried across the Atlantic in ships which provided a base for catching and for drying ashore. This fishery was called the sedentary fishery because the ships remained in the same harbour or inlet where the fish was dried on the beaches. The dried cod was exported to Mediterranean countries and to the Antilles.

Two types of vessel were used in the sixteenth and seventeenth centuries. The first was of Dutch origin with broad sides, flat decks and rounded shape. They were heavy ships with large holds which could ground on the sand without danger; hoys had one mast and a spritsail, hookers had two masts, a mizzen, bowsprit and spritsail, and a dogger was a two-masted vessel developed for use on the Dogger Bank in the North Sea. The second type was of Portuguese and Biscayan origin. Such vessels were longer with finer hulls with rising decks and square sterns. They had two or three masts, were fast and manoeuvred easily and hence could work well off rocky shores. Both types of vessel were built in France in as many as fifty ports. In the seventeenth century, galliots, hoys and frigates were used as cargo vessels. Pinnases were small ships with square poops, three masts and square sterns, of about 150 to 300 tonnes. In the eighteenth century, three masted vessels were in use, but in 1720 the brigantine and schooner with Bermuda rig of American origin were intro-

Figure 29. The French cod fishermen at work on the Grand Bank (above); the hook and line fishermen worked alongside each other on deck; the splitters were dressed in leather aprons and they stood in barrels (Duhamel du Monceau 1769).

duced to the French fishery; they were manoeuvrable and useful in the salt cod fishery. In the sixteenth century, the vessels were small, of 40–100 tonnes; in the seventeenth century, vessels of 40–100 tonnes were used in the salt cod fishery but in the dried cod fishery vessels of up to 250 tonnes were employed; in the eighteenth century larger vessels of 200–300 tonnes were built.

The salt cod fishery

The salt cod fishery took place on the Grand Bank, St Pierre Bank, Green Bank, Banquereau and Orphelin Bank. The captains steered by latitude

Figure 30. The salting of the cod. The split fish were roused in salt and laid out to dry (Duhamel du Monceau 1769).

until they were 'banqués' in 60–100 m, where they started to fish. On passage the fishermen built a fishing gallery for line fishing. Each man stood in a barrel against the weather, with straw around the top to protect him from the hooks; there was a false bottom to let out the water. The fisherman wore a large leather apron up to the chin, which kept out the water (Figure 29). The barrel was lashed to the deck. The lines were 160 m in length and each man had eight to twelve. There was one hook on each line and they differed according to the size of cod; they were made of iron because the sea bed was stony. A little piece of copper covered the tail of the hook to protect the hands of the fishermen. There were various forms of hook, some baitless, some steel, some serrated; the scythe hook was made of two hooks laid back-to-back for jigging. English hooks were smaller, but they were worked from an anchored ship, whereas the Frenchmen always drifted.

They drifted with the wind on the starboard, with little sail. The lines were first baited with salt herring, then later with cod guts. Each line was weighted with a 1.8–2.3 kg lead. As the vessel drifted, the line was hauled a little every now and then. When the line was hauled, the lead was pulled in first; then the fish was gutted, the tongue cut out, and the hook was rebaited with whatever was in the cod gut. The fish was then thrown in the pound behind the fisherman. Behind the fishermen, the header worked at a table and the dresser cut the fish into pieces and threw them into the hold through a hole in the table; both stood in barrels and wore leather aprons (Figure 29). Below, the salter, dressed in canvas, laid the fish in salt for twenty-four to forty-eight hours. Roes were kept for the crew, cod liver oil was used for lamps and air bladders were converted to glue. The ship's boy counted the fish and the fishermen with least catch cleaned the pound. The crew caught about 100–120 cod each day; between 16 April and 30 September the *Lorier* of Honfleur with a crew of twelve men caught 16670 cod. The French anchored at night; the English sailed at night but they had bells and they shouted in the fog.

The crew comprised captain, pilot, boatswain, carpenter, salter, header, ten sailors, five novices and two boys. After 1681, a surgeon and a priest were obligatory. Victuals were carried for six months, biscuit, peas, beef, lard, bread, olive oil, chickens, eggs, dried cod, butter, wine and brandy. Salt was available in Brittany, Poitou and Lower Normandy, but it was heavily taxed; in the sixteenth century, salt was imported from Spain and Portugal, but in the seventeenth century it was taken from the Ile de Ré. At the end of each voyage, the remaining salt was returned to France. After 1743, strong drink was forbidden by royal authority because of drunken mutinies.

The dried cod fishery

The ships were larger than those in the salt cod fishery, up to 300 tonnes and they carried 150 men (ten dressers, ten cutters, sixty fishermen, twenty stagemen, carpenters and caulkers, eight to ten men for the capelin boats, eight to forty boys, a priest and two surgeons). Up to seventeen longboats (chaloupes) were carried and each had five men, three on the boat and two on the drying stages; the boat's crew comprised the boatmaster, the bowman and the banksman. The boats left at dawn to go to the grounds at a distance of 15–20 km; each had a barrel of water, or small beer, and some bread. From each harbour, thirty or fifty boats would set sail and each master was free to go where he liked to anchor and fish. Each of the three men had a line on each side of the boat with hooks about 2 m off the sea bed. Mackerel was the preferred bait, but herring and capelin were also used; indeed two or three boats were devoted to a subsidiary capelin fishery, particularly in June and July. Small seines were used for this purpose. The boats returned between four and six in the evening and moored at the end of the stage.

The stage was 24–30 m long and 6 m wide; one end was at the high-tide mark and the other at the bollards where the boats tied up (Figure 31). At one end of the stage stood the splitter and at the other the cutter; three cuts were made, above and below the backbone and from the third the backbone was excised (Figure 29). In the middle of the table there was a ring for salt, 6–7 m long and 1.2 m wide. The salter laid the cod head to tail, skin beneath and flesh on top. After salting, the fish were washed in cages in the sea and were dried on a long beach. They were dried for ten 'suns' and turned over every day; a 'sun' was a period of some days which varied somewhat during the summer (Figure 32). Along the beach 5 m were allocated to each boat. As many as twenty ships might have anchored in the harbour or inlet, so the drying beach might have been as much as 2 km in length. If the beach was gravelly it was cleaned, and if it was sandy it was covered with branches of spruce. Not far from the stage stood the two floor cabin, half covered with a sail. The sides were garnished with branches of spruce. There were two floors and men slept two by two (separated by poles) on palliasses filled with dried grass.

Because the ships carried large crews on a voyage of six months or so, large amounts of victuals were carried. Such a vessel would carry thirty large casks of biscuits and twenty-six large casks of wine. In the harbour, the crew fed on eggs, chickens, sausages, oysters, sardines, cheeses, grapes, figs, onions and of course a soup of cod. They drank tea, coffee, wine and brandy; drinks were needed in the warm foggy weather. For medicaments, treacle, hyacinth, scurvy grass and woundwort were carried.

Figure 31. The stages on which the cod were landed, cleaned, sun dried and salted; the dory at the foot of the stage is the catching vessel and the ship offshore is the transatlantic carrier (Duhamel du Monceau, 1769).

Figure 32. Cod were dried in the sun for long periods (Duhamel du Monceau 1769).

The main fishery lasted from June to mid August, but the vessels did not leave until the end of September so drying could be finished. The fish were laid on branches in the hold and the boats were left behind; the establishments were quite extensive and frequently captains returned year after year to the same anchorage. However, the establishments used by the sedentary fishery were the source of three forms of conflict: (*a*) between ships' captains on who owned what and frequently codes of conduct were agreed between them; (*b*) depredations by Eskimos and Indians, particularly in Labrador; (*c*) with English vessels, particularly during the eighteenth century.

The disposal of fish

The major ports for salt cod were Dieppe, Le Havre, Honfleur, St Malo, Granville and Sables d'Olonne. In the salt cod landings, the fish were sorted by size and those less than 36 cm in length were discarded. Sorters in three bands of two men each met the incoming vessel. There were four main categories: (1) gaffe cod, (2) officer's cod, (3) market cod, (4) sorted cod. The first two were of 1–2 m in length; market cod were 65 cm in length and the sorted cod were of 51–65 cm in length. In 1740 it was said that the length distributions had remained the same for a hundred years. There were four other categories: chafed cod, valid cod, corrupt cod (thin or badly salted) and rejected cod. So the sorters checked size and quality and the tally keeper monitored the sale. The salt cod was consumed in France, mainly in Paris, which is why Dieppe, Le Havre and Honfleur were ports of major importance. But the region between Nantes and Bordeaux was supplied from Sables d'Olonne.

The ports for dried cod were Bordeaux, Nantes, La Rochelle, Marseille, Granville, St Malo and St Brieuc. Two vessels landed 25 and 60 tonnes, respectively; in 1770, *l'Heureuse Marie* landed 495000 cod, a much larger quantity. There were three main sale categories, 'great market', 'medium market' and 'small market'; rejected categories were broken, oily, scorched, blotched, badly cut, hard and burnt. The dried fish were exported to Mediterranean countries and later directly to the Antilles or the West Indies.

History of the fishery

In the fifteenth century, cod was taken from the North Sea and Iceland; the English dried it but the French did not. By 1504, Portuguese vessels sailed for the Grand Bank and by 1550, 150 ships sailed from the port of Aveiro and they worked between Bonavista and Cape Race. The French fishery probably started when eleven Norman vessels appeared in St John's in 1527. The fishery between 1510 and 1540 was primarily for salted cod and the dried cod fishery did not start until the second half of that century. Basques worked on the west coast of Newfoundland and the Malouins (from St Malo) in the Strait of Belle Isle.

The sedentary fishery for dried cod was based on the Petit Nord (the eastern coast between Bonavista Bay and Cap Normand), Placentia and at Gaspe and l'Ile Percée; beaches and stages were allocated in order of arrival. Sabine (1853) gives figures of 150 vessels in 1577 and 1578 and 100 in 1615. Eskimos from Labrador sometimes opposed the operations, so the Malouins armed their convoys from 1636 onward (in 1687 one vessel had twelve cannon and the crew were armed with pikes and eighty muskets). On the Gaspé peninsula and as far east as Miramichi, the fishermen came from Bayonne, St Jean de Luz, Le Havre, Honfleur and La Rochelle. At Miramichi and l'Ile Percée some fishermen overwintered in order to preserve the stages, but, in 1686, Governor de Meulles ruled that vessels from metropolitan France had preference over those of the inhabitants. The winterers tended to fish in spring and autumn and the metropolitan fishermen worked in summer.

In the seventeenth century, the French abandoned Cap Breton and Labrador and concentrated sedentary fisheries on the Petit Nord, la Baie des Chaleurs and Acadie (Nova Scotia). Colonies were established in Nova Scotia and at Placentia. During the War of the Grand Alliance (1689–97), Placentia was attacked by the English on three occasions and the French destroyed the English colony at St Johns. In the Treaty of Utrecht after the War of Spanish Succession (1702–13), France ceded Newfoundland and Nova Scotia to England. During this period, Moroccan pirates in the Strait of Gibraltar awaited the dried cod vessels en route for Marseille, so fishing vessels were sometimes armed and frequently sailed in convoy.

In the eighteenth century, the main French base was at Cap Breton and, because the grounds were at some distance, the nature of the fishery changed. In 1720–3, the *goëlette* was introduced; they displaced 40–60 tonnes and carried ten to eleven men, only three or four of whom were sailors. Sabine (1853) gives the number of vessels as 400 in 1721, 564 in 1744 and 100 in 1745. Such vessels sailed 10 to 100 km to Sable Bank, St Pierre Bank and Green Bank. The salted cod was dressed aboard; when the hold was full, the vessel returned to its Newfoundland harbour and fish were dried ashore. In 1778 another new method was developed; a long line set with a hook every 2 m was laid on the sea bed in the evening and it was hauled at dawn on the following day. The method came into general use in the last decade before the revolution in 1789.

After the Seven Years' War (1756–63), in the Treaty of Paris, France was free to fish and establish drying establishments on parts of the coast of Newfoundland. The French fishermen were limited to 15 km from British coasts in the Gulf of St Lawrence, but 75 km from Cap Breton, Nova Scotia and elsewhere. A sedentary fishery was established on St Pierre et Miquelon. The vessels were mainly *goëlettes* and brigantines.

De la Morandière (1962–6) reported the number of vessels in the eighteenth century in twenty-four ports for fifty-three years of the century. For two of the larger ports, Granville and St Malo, the distribution of vessels in the dried cod, or sedentary, fishery and the salt cod, or Bank fishery, 47.6% in the former and 52.4% in the latter. In Table 1 are given numbers of vessels, boats and catches in certain years.

There is conflict between the two sets of figures from 1786–9; I assume that the number of vessels are underestimated but that the number of boats may be correct. The average number from the French Colonial Office was 320, not very different from the return of 352 from the Inventory of 1664.

In the sedentary fishery at Cap Breton, de la Morandière gives catches per vessel in quintals:

Year	Quintals[a]
1720	1800 (417)
1721	2352 (544)
1726	2400 (556)
1729	1878 (435)
1733	3127 (724)
1735	3163 (732)
1736	3733 (864)
1738	3233 (748)
1739	2852 (660)
1742	3276 (758)
1743	3471 (804)

[a]Numbers in parentheses are tonnes.

After the Napoleonic wars the French retained the islands of St Pierre et Miquelon and the right to work the sedentary fishery from the French shore. The vessels used were *goëlettes* for the sedentary fishery and for the salt cod offshore. In 1817, some fishermen were still using handlines, but the more usual gear was the long line. At the end of the *ancien régime*, such lines were 600–800 m in length but in the nineteenth century lines of up to 6000 or 8000 m long were used. From the longboat, 35 pieces of 120 m (about 4 km total) each were shot from the starboard side and 20 pieces from the portside (about 2 km); a hook was set every metre. Longboats were used until 1870 when they were replaced by dories – light, flat-bottomed boats of American origin; they were stacked inside each other. They were manned by two men. Cotton lines were used, ten times as long as the earlier hemp ones; small steel hooks were used. Hence, the new gear was lighter and much more effective. Dories caught twice as much as the longboats.

Before the revolution the pickled cod fishery played a secondary role to the sedentary fishery for dried cod. As the century progressed, the big three-masters sailed fully equipped directly to the Grand Bank. When they arrived they furled the white sails used for the crossing and replaced them with tan sails, preserved against the fog. In 1890, the French were denied the right to buy bait in Newfoundland and they turned to a whelk which lives on the Banks. In 1904 the French relinquished their ancient rights in Newfoundland, the French shore and their sedentary fishery died out. Before the First World War, there were two or three fisheries

The provident sea

Table 1. *Catches of cod and number of vessels in the eighteenth century*

Year	Number of vessels[a] A	B	Number of boats	Quintals cured[b]	
1764	240				
1765	249				
1767	381				
1768	384				
1769		431	1455	215030	(49775)
1770	287	437	1470	455340	(100773)
1771		419	1327	239864	(55524)
1772		330	1468	388800	(90000)
1773	264	284	1452	336250	(77836)
1774	274	273	1614	386215	(89402)
1777					
1784	328				
1785	358				
1786	386	86	1532	426400	(98703)
1787	347	73	1342	128590	(29766)
1788	346	86	1560	241262	(55848)
1789	313	58	1035	239000	(55324)
1791		43	628	40580	(9394)
1792		46	689	94000	(21759)

[a]Number of vessels (col. A) from French Colonial Archives quoted by de la Morandière (1962–6).
Number of vessels (col. B) and number of boats and quintals cured from Lambert (1975), based on the returns of British admirals. (Sabine (1853) gives another set of figures for this period, which gives a rather lower number of vessels.)
[b]Numbers in parentheses are tonnes cured: 4.32 quintals represent 1 tonne of fresh fish (Lambert 1975).

during the year. Between the end of March and June, they would work the Banks and then sail to St Pierre to disembark the fish and revictual. After 1918, the vessels stayed at sea all the time. Their lines were of 24 pieces, of 110 m each and the last of 133 m in length. The long line was thus 2 km long with 75 hooks. It was anchored and buoyed, moored to a long hawser called *la sabaille*. It was shot each night and hauled next morning.

Towards the end of the age of sail, the *goëlettes* displaced 350 to 450 tonnes, carried a crew of thirty men and ten to twelve dories. After the First World War there was a fair number of such large vessels. By

Table 2. *Catches of cod and number of vessels in the American fishery in the early nineteenth century*

Year	Number of vessels	Number of men	Quintals of fish[a]	
1816		8108		
1823	184	3655		
1824	348	6672		
1825	336	6311		
1826	341	7088		
1827	387	8238		
1828	381	7957		
1829	414	9428		
1830	377	8174		
1831	302	6342	300000	(69444)
1833		10000		
1834		10000		
1835			300000	(69444)
1839		11499		
1841	400	11900		
1843	400			
1847		12000	450000	(104166)

[a] Figures in parentheses are in tonnes.

1931, there were only four sailing vessels from Granville and by 1938, 19 from Fécamp. The sedentary fishery declined for three reasons during the nineteenth century: (a) the French shore became colonized, (b) it became difficult to recruit labour for the stages in the hinterlands of the home ports in France, (c) the cod fishery at Iceland increased.

De la Morandière lists the number of vessels by port in certain years during the nineteenth century; the major ports were Fécamp, Granville, St Malo and St Brieuc. The number from Fécamp increased from ten in 1825 to between forty and fifty between 1845 and 1885 and remained at that level until the First World War. Numbers from Granville increased after the Napoleonic wars to eighty until about 1860, after which numbers declined to thirty-six in 1885 and twenty-four in 1914. Seventy vessels sailed from St Malo in 1821 and numbers remained as high as ninety up to 1865, after which they declined to fifty-seven in 1890. From St Brieuc there were forty vessels from 1820 to 1870, seventy vessels between 1850 and 1865 and subsequently numbers declined to four in 1900. After the Napoleonic wars, the French fleet built up to about 200 vessels in mid century, after which numbers declined rather slowly to the beginning of the Second World War. Figures relating to this by Sabine (1853) are given in Table 2.

Prowse (1896) quoted the following quantities of French exports of dried cod from Newfoundland:

Year	Quintals	Tonnes	Year	Quintals	Tonnes
1878	201982	46755	1884	404604	93658
1879	233923	54149	1885	497284	115112
1880	254939	59013	1886	579390	134118
1881	236250	54687	1887	453058	104875
1882	255671	59183	1888	338126	78270
1883	304580	70505	1889	300000	69444

The average from Prowse's figures is 338317 dry quintals (nearly 76000 tonnes of fish fresh from the sea); Prowse quoted the figures to show the decline in 1888 after the Bait Act was passed, preventing the French from catching bait in the inshore waters of Newfoundland. It might be reasonable to guess that the catch at mid century amounted to 88000 tonnes after which it declined to 68900 tonnes and 49200 tonnes by 1904.

The British fishery (including the English from West Country ports and that from Newfoundland)

English vessels visited Newfoundland before 1550; 'Newland' is recorded in Acts of Henry VIII and Edward VI (Lounsbury 1934). But the English fishery did not develop until the last two decades of the century; Parkhurst's letter (1578, quoted by Lounsbury 1934) recorded 150 French sail, 50 Portuguese, >100 Spanish and 50 English. In 1583, Sir Humphrey Gilbert claimed Newfoundland for the Crown. In 1580, the Danes had introduced licence fees for the Icelandic fishery and the English tended to shift ground to the Grand Bank. They bought salt from the Portuguese and, by 1594, 100 vessels sailed from West coast ports. During the war with Spain (1584–4), Spanish and French vessels were captured by English boats; in 1591, Raleigh took 108000 dried fish, 4000 green (salted) fish and 14 hogsheads of train oil (cod liver oil). In 1597, Dutch, Irish and French boats visited Plymouth in September to buy fish from the incoming Newfoundland fleet. At the same time, ships from London carried dried cod to the Mediterranean countries (Lounsbury 1934). In the year 1600, there were 200 vessels, and 250 in 1605 (Judah 1933).

Lewes Roberts (quoted by Sabine 1853) wrote that there were 500 vessels in Newfoundland in the early seventeenth century. The fleet included sack ships to take fish to the Mediterranean and they filled their holds with sherry for the West Country ports. Newfoundland was also a half way house to New England, Virginia and the West Indies.

In 1610 the London and Bristol Company was chartered for trade and plantation in Newfoundland; their merchants tended to send fish directly to markets in south-west Europe. English, French, Flemish, Dutch and German vessels sailed directly to Newfoundland, excluding the West Country ports from which most of the English fishermen came. The conflict between the London merchants and the West Country fishermen continued for the best part of two centuries. The first Western Charter of 1634 laid down eleven rules for trade and fishery, recognized 'admirals' (captains who arrived first); they were to be enforced by the mayors of the Western Ports and the vice admirals of Hampshire, Dorset, Devon and Cornwall. A second charter was issued in 1661 and a third in 1676, but the argument continued.

A fishery also developed from New England. In 1603, George Waymouth reported that some cod from Newfoundland were 5 feet long and 1.5 m round. In 1645, Boston vessels visited Bay Bulls in Newfoundland; there was a market in Spain for the products of the winter fishery and another market for poor-grade fish later for slaves in Virginia and the West Indies. Dried fish were sent in 1650 from Newfoundland to Boston and thence to Lisbon, Marseille, Bordeaux and Toulon.

A vessel of 100 tonnes carried forty men (twenty-four fishermen, seven headers and splitters, two boys to lay fish on the table, three to salt fish and three to pitch salt on land and to wash and dry the fish). Such a vessel would carry nets, leads, hooks, lines, bread, beer, beef and pork, which supported employment of many bakers, brewers, coopers, carpenters, smiths, net makers, rope makers, line makers, hook makers and pulley makers in the West Country. Salt was bought from Spain, Portugal and France. A vessel of this type would bring back 2200 quintals of dried fish (roughly 500 tonnes). In 1617, Mason, in his *Brief Discourse*, recorded that there were 'cods so thicke by the shoare that we nearlie have been able to row through them. I have killed of them with a pike' (see Sabine 1853). The West Country fishermen sailed in mid March and arrived in Newfoundland towards the end of May. The first two or three weeks were spent repairing the stages. Shore seines were used to catch capelin for bait. The fishery ended in the third week of August and the dried fish was packed aboard the vessel which sailed before the autumn storms. A by-product of the fishery was the cod liver oil, or train oil, rotted down, cleared and poured into hogsheads.

Between 1615 and 1621 there were 300 English vessels in the harbour at St John's. In 1618, Whitbourne reported that 5000 men sailed in 250 vessels from West Country ports and caught 30 million cod (Judah 1933). Sir William Vaughan wrote that 300–400 English vessels sailed yearly to Newfoundland; in each boat, three men would catch 25 000 to 30 000 fish in thirty days. From 1624 the English trade declined, partly because of wars with France and Spain but also because of attacks by pirates. Indeed in 1612, Captain Peter Easton and ten warships took one fifth of the men and provisions of the English fleet. There were Barbary vessels, but also there were the Sallee rovers who were French, Spanish and English pirates under the Turkish flag and, on one occasion at least, they actually raided West Country ports. By 1634, the trade recovered somewhat and there were 270 vessels in Newfoundland from West Country ports. But in 1638, Sir David Kirke was appointed governor, for there were 500 residents between Cape Bonavista and Trepassey.

Between 1650 and the end of the seventeenth century, the English fishery from West Country ports suffered from three distinct sources, the conflict with the inhabitants of Newfoundland, the competition with France and the competition with New England. The wars with Spain (1656–9) and with the Dutch (1672–4) militated against the English fishermen; in the first 1000 fishing vessels were lost. Sir David Kirk had started the bye-boatmen, passengers on the English vessels who fished independently and sold their catches directly to the sack ships. In 1671, the Council for Foreign Plantations concluded that the English fishery should be protected and 'masters of ships be required to bring back all seamen, fishermen and others and none to be suffered to remain in Newfoundland . . . inhabitants of Newfoundland were to be encouraged to Jamaica and other foreign plantations' (Lounsbury 1934). In 1676 the convoys were armed. The French fortified Placentia in 1662 and Colbert excluded the English from the French dried-fish market; in the Treaty of the Pyrénées, the French obtained entry to the Spanish dried-fish trade. The New Englanders entered the market with cod caught on George's Bank and exported it directly to Spain and Portugal. By 1679, they had established a three-cornered trade, rum and molasses from the West Indies, wine and brandy from the sack ships and fish and tackle from the Newfoundlanders. In the Navigation Act of 1651, foreign-produced fish and train oil were not to enter England; the Newfoundland Act of 1663 relieved the tax on cod caught by Englishmen and Newfoundlanders. But such measures did little to modify the effects of competition.

Towards the end of the seventeenth century, war broke out in Europe. Between 1689 and 1697 there was the war of the Grand Alliance between the French on one hand and the English and Dutch on the other. During this period the French made various attacks on the English settlements.

In the Newfoundland Act of 1699, regulations rather similar to those of the Charters of 1634, 1661 and 1676 were passed, but there were no restrictions on the planters or the bye-boatmen; the fishing admirals were made responsible but did not enforce the regulations. After the War of Spanish Succession (1701–14), the Treaty of Utrecht was signed; the French ceded Nova Scotia (except Cap Breton and Prince Edward Is.) and the south coast of Newfoundland including the fort at Placentia. However, they remained free to fish on the 'French shore' (the western coast) and on the 'Petit Nord' (the eastern coast).

In 1700, the distribution of English vessels sailing to Newfoundland by port of origin was: Bideford, 30; Barnstaple, 12; Plymouth, 12; Bristol, 6; Dartmouth, 3; Jersey, 4; Guernsey, 5; Topsham, 25; Weymouth, 3; Lymington, 1; Portsmouth, 1; Poole, 11; Southampton, 2; Liverpool, 4; Lynne, 2; Dublin, 1. They were distributed mainly in the West Country and the fish caught were exported to the Mediterranean by the sack ships.

In the early eighteenth century, the Newfoundlanders developed a new boat in the fishery, the shallop. It was 9–12 m long in the keel, decked fore and aft where the fishermen could sleep. It was very beamy and carried five men for five or six days at sea on the Grand Bank; they caught 200 quintals per day as opposed to 150 quintals per day in the inshore fishery (1 quintal = 0.22 tonne). The fishermen were dressed in whitney and barked swanskin. Bigger vessels, such as barks would take up to 600 quintals per day. This method was taken up by the West Country fishermen and the admirals became so busy on the Bank that they could not regulate the dry fishery. As civil government was established in 1729, the efforts of West Country interests against the Newfoundlanders died away.

There were three wars in the eighteenth century of importance to the Newfoundland fishery, the War of the Austrian Succession (1742–8), the Seven Years' War (1756–63) and the American War of Independence (1776–83). During the Seven Years' War, the English fishermen entered the Petit Nord, the eastern coast north of Cape Bonavista. In 1775, Lord North (quoted by Lounsbury (1934)) put forward a Bill, 'to restrain the trade and commerce of the provinces of Massachusetts Bay and New Hampshire and colonies of Connecticut and Rhode Island and Providence Plantation, in North America to Great Britain, Ireland and the British Islands in the West Indies; and to prohibit such provinces and colonies from carrying on any fishing on the banks of Newfoundland, or other places therein mentioned . . . ' The purpose was to restrain the competition from New England. Article III of the Treaty of Versailles (after the American War of Independence) stated: 'the United States shall continue to enjoy unmolested the right to take fish of every kind on

The provident sea

the Grand Bank and on the other banks of Newfoundland; also in the Gulph of St Lawrence and at all other places in the sea where inhabitants of both countries used at any time to fish.' Thus a notable conflict was resolved.

Table 3 shows the number of vessels, boats and catches in quintals. The vessels are separated into fishing vessels, sack vessels and colonial vessels (from New England); the boats are separated into British, bye-boats and Newfoundland, as are the catches. Also the number of stages is given. The material is arranged by years intermittently between 1615 and 1792 (Lambert 1975; Lounsbury 1934; Prowse 1896). The fleet of English vessels amounted to about 200 in the early seventeenth century, after which it declined to low levels in the 1780s; apart from the years 1698 and 1700, numbers were low during the War of the Spanish Succession. After the Treaty of Utrecht in 1713, numbers increased to over 300 before the American War of Independence; after that war the fleet was reduced a little. The number of sack ships followed roughly the same course, as might be expected. The numbers of colonial (i.e. New England vessels) visiting Newfoundland increased through the eighteenth century until 1774, but they were much reduced after the American War of Independence. The distribution of boats is very interesting, showing first the development of the bye-boatmen during the eighteenth century, but more dramatically the increase in boats belonging to Newfoundlanders,

Table 3. *Numbers of vessels (Fishing, Sack and Colonial), boats (British, bye-boats and Newfoundlander), catches in quintals (British, bye-boats and Newfoundlander) and number of stages*

	Vessels			Boats				
Year	Fishing	Sack	Colonial	British	Bye- boats	Newf.	Catch, quintals	Stages
1615	170							
1644	270			2160			540000	
1660								
1672	23							
1675	175			677		277		
1676	126			894		206		
1677	109			892		337		
1680	97							
1682	32			183		299		
1684	43			294		304	265198	
1698	143			532		397		

Table 3 (*cont.*)

Year	Vessels			Boats			Catch, quintals	Stages
	Fishing	Sack	Colonial	British	Bye-boats	Newf.		
1699		236		805	115	467		465
1700	172	49		800	90	674		583
1701	75	46		338	97	558	216320	544
1702	16			35		380		
1703	23			44		214		
1704	23							
1705	20			60		200		
1706	46			136		232		
1707	70			196		257		
1708	49			170		356		
1709	35			130		258		
1710	49			153		365		
1711	62			168		439		
1712	66			198		370		
1713	46			162	195	408		
1714	85	45	20	380	160	500	115000	450
1715	108	38	42	376	197	464	89622	440
1716	86	30	31	319	184	408	88469	376
1749	80	125	66	171	349	654	506406	725
1750	93	115	75	199	485	748	432240	785
1751	122	87	103	295	542	668	358310	835
1764	141	97	205	210	366	1236	561310	994
1765	177	116	104	318	345	1156	532512	1005
1766	204	104	83	536	361	1117	559985	1039
1767	258	92	115	490	372	1151	553310	1138
1768	296	93	114	472	437	1195	573450	1208
1769	354	117	120	430	429	1333	578624	1187
1770	368	123	138	528	444	1229	649498	1144
1771	369	120	123	556	559	1173	644919	1163
1772	306	146	138	490	605	1330	759843	1132
1773	262	93	125	479	560	1276	780328	1167
1774	254	149	175	451	518	1446	695866	1219
1784	236	60	50	572	344	1068	437316	942
1785	292	85	58	424	540	1434	545940	1135
1786	280	173	34	276	413	1152	569142	1170
1787	306	167	37	321	451	1709	732015	1444
1788	389	150	28	273	317	2090	949950	1578
1789	304	168	70	413	533	1456	771589	1464
1790	259	143	69	370	387	1414	649092	1334
1791	245	151	76	375	584	1259	536289	1380
1792	276	161	57	250	1997		552260	2356

Table 4. *Export of cod from Newfoundland in the first two decades of the nineteenth century in quintals*

Year	Spain, Portugal and Italy	Great Britain	West Indies	Canada	United States	Brazil	Total
1804	354661	189320	55998	18167	43131		661277
1805	377293	65979	81488	22776	77983		625519
1806	433918	84241	100936	32555	116159		767809
1807	262366	130400	103418	23541	155085		674810
1808	154069	208254	115677	40874	56658		575532
1809	326781	292068	133359	41894	16119		810219
1810							884470
1811	611960	139561	152184	18621	1214		923540
1812	545451	67020	91867	4121		2600	711059
1813	706939	50678	119354	14389			891360
1814	768010	55721	97249	24712		2049	947741
1815	952116	46116	159233	24608	588		1182661
1816	770693	59341	167603	37443	2545		1037625

From Innis (1940).

by a factor of 3 or 4 during the century. In the early seventeenth century, catches may have been as high as half a million quintals of dried cod. During the eighteenth century total catches were greater in the years just before the American War of Independence; after that war, catches recovered to high levels. In the latter half of the century, the British catch amounted to one third and one half of the total, the rest being taken by the Newfoundlanders and the bye-boatmen. The number of stages increased by a factor of 4 during the century. The catch of half a million quintals of dried cod corresponds to one of about 100000 tonnes, round weight (see below for the conversion).

During the latter part of the eighteenth century, from 1769 onward, the proportion of English vessels working on the Bank (and not in the inshore region) comprised about half the total number. In the early 1790, the Norwegian stockfish caught at Iceland became cheaper, and in 1785 and 1792 the duties in Spain and Portugal were increased. The Newfoundland exports of dried cod in quintals during the Napoleonic wars and just after are detailed in Table 4.

In 1806, the Newfoundland bye-boat fishery ceased and the gangs impeded the English fishery. The seal fishery started in 1799 when the ice carried the animals far south. In that year, 120000 seals were captured and, by 1804, 156000 were taken (see Chapter 8). This fishery and the increase in the ownership of vessels contributed considerably to the decline of West Country ports. Norwegian competition had increased and by 1827, sixteen million cod were caught in the Vestfjord. The Petit Nord was revisited by the French. By the beginning of the 1830s not more than eight or ten British vessels were employed in the Bank fishery.

Figure 33 shows the catches of cod in Newfoundland in quintals (dry) from 1805 to 1935, together with catches from the 'bankers' from 1889 to 1934 (Kent, Watson and Little 1937). McPherson (1935) gives conversion factors for quintals for wet fish from the sea (before heading and gutting) to dressed dried fish, 4.47 for the shore cure and 3.77 for the more heavily salted Labrador cure. There are 112 lb (51 kg) per quintal, so the conversion factors to tonnes of fish from the sea is 0.2271 for the shore cure and 0.1915 for the Labrador cure. Kent *et al.* (1937) described the methods of fishing in the 1930s and the use of the hook and line is obviously much like that in earlier centuries, if the gear was much lighter. In Figure 33 the conversion used is 0.2271 because the Labrador cure must have comprised a small proportion. In general the Newfoundland industry increased its output from 750000 quintals (about 172500 tonnes of fish from the sea) to perhaps 1500000 quintals on average (about 345000 tonnes of fish from the sea) during the first thirty years of the

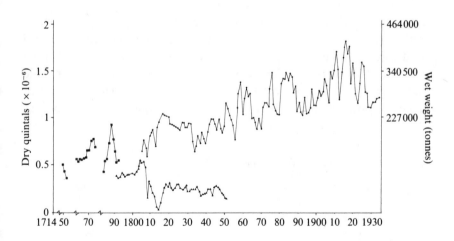

Figure 33. Recorded catches from 1714 to 1935: catches from the sedentary fishermen and the bankers, 1805–35, in quintals (dry weight) and in tonnes (wet weight) (●); catches on the Grand and Western Banks, the north-east shore and on George's Bank (▲) (Anon. 1937). For an explanation of the quintal conversion, see the text.

twentieth century. The 'bankers' caught a much smaller quantity, about 100000 quintals, or 22700 tonnes of fish from the sea.

In the Convention of 1818 (after the 1812 war between Britain and the United States), US fishermen became free to take fish of all kinds on the southern coast of Newfoundland and to dry and cure fish in unsettled bays and creeks. In 1904, the French withdrew from the French shore, after years of dispute as to whether they should catch bait (which could be eaten by cod) and whether the French should maintain the stages.

The North American fishery

New England developed on the basis of its fisheries. In 1675, 665 vessels caught 350000 to 400000 quintals. by 1731, there were 230000 dry quintals for export. In 1721, 20000 quintals of cod were cured at Canso on Cap Breton. The schooner had replaced the shallop (or *chaloupe*) in 1713, and in 1741 seventy schooners from Gloucester visited the Banks to 'go on their own hooks'. By 1747, the exports from New England ports to Mediterranean countries reached 300000 dry quintals. There are two points of interest in the history of the Newfoundland cod fishery: (*a*) the export of dried cod, caught first on George's Bank and off the coast of New England, directly to the Mediterranean; (*b*) the capture of cod on the Grand Bank and other banks off Newfoundland.

The Treaty of Paris in 1783 recognized the rights of American fishermen to catch fish on the high seas and granted the privilege of fishing within British jurisdiction under certain conditions. But a British Act and Order in Council in the same year restricted the trade between the United States and British colonies to British ships and prohibited American trade in fish with the British West Indies; in response, the Americans introduced various bounties. Their exports (in dry quintals) developed as follows:

Year	To Europe	To West Indies	Total
1786–90	250650	142050	392700
1800	144353	244353	392726
1790–1808			438453
1809–18			200437

Between 1790 and 1810, an average number of 584 vessels visited the Grand Banks; each made three trips per year and landed 510000 quintals each year. To the Baie des Chaleurs and Labrador, 648 vessels sailed and landed on average 648000 quintals each year, a total of 1 158000 quintals

(McFarland 1911). Perhaps the export figures given above are incomplete. The Convention of 1818 granted fishing rights to American fishermen on the southern and western coasts of Newfoundland and in the waters about Magdalen Is., and the facilities for drying and curing fish were enlarged.

From 1885 onward, both fisheries on the Grand Bank (and the Western Bank) and on the New England banks declined. McFarland wrote that the period 1866–85, after the Civil War was one of general prosperity, but it was less prosperous in fact than the period 1845–65. The subsequent decline was due partly to increased activity in Canada and Newfoundland in the export to Mediterranean countries and partly to competition within the United States.

Conclusion

The fishery for cod on the Grand Banks started in the early sixteenth century and it persists today. Records are necessarily patchy and not until the eighteenth century can we estimate catches or number of ships. In earlier centuries, there are some records which give some indication that the effective effort was considerable. From isolated reports, 150 French vessels in the sixteenth century visited Newfoundland each year and perhaps 300 from 1600 to 1750. There were possibly 250 visiting English vessels in the first half of the seventeenth century and about 100 in the second half. At an average catch of 2000 quintals vessel^{-1}, up to 250000 tonnes of fish fresh from the sea were taken from about 1580 to 1750. A rough sum of intermittent catch records between 1750 and 1900 suggests that in the first fifty years about 800000 quintals were caught. Between 1800 and 1850 perhaps more than two million quintals were landed and in the second half of the century the quantity declined to 1.74 million quintals. Perhaps as much as 400000 to 500000 tonnes of cod were taken from the region of the Grand Bank during the nineteenth century.

In the early stages of the fishery, the fish were quite large. Between 1640 and 1740, the market cod were 65 cm in length and the 'gaffe' cod were 1–2 m in length which is large for a size category.

The methods of capture remained much the same in the sedentary fishery for over 400 years. In 1720–3, the French introduced the larger *goëlette* and in 1778 the long line. In the same century the Newfoundlanders devised the shallop, a sea-going long boat. The Americans introduced the dory, and lightened the hooks and lines. But the stages and the essential method of hook-and-line fishing remained the same throughout the whole period up to the Second World War.

The demand for *bacalão*, the dried cod, in Catholic countries was strong enough to sustain the largest and most persistent preindustrial fishery. As it declined in the first three decades of the present century trawlers started to work on the Banks. Today the main fishery off Newfoundland is prosecuted by trawlers. They are large, safe and comfortable vessels in sharp contrast to the small boats of about 50 tonnes burden which crossed the Atlantic twice a year in the equinoctial seasons.

4

The North Sea herring fishery

Le Hareng est une de ces productions dont l'emploi décide de la destinée des Empires. (Lacépède 1841)

The early history of the herring fishery comprises scattered laws regulating disputes and records of rents, taxes and purchases. Salt herring were used in Lent and on winter fast days and by soldiers and travellers. Packaged food in barrels was carried across Europe in wagons throughout the Middle Ages. The fish were caught by drift nets and were salted and pickled ashore until the Dutchman William Beukelsen invented a method of pickling and curing at sea. Salt was available quite close to each of the ports in the three early fisheries, near Yarmouth on the east coast of England, near Falsterbö and Skanör in southern Sweden and near Bievliet in Holland.

The Scanian fishery lasted from the eleventh or twelfth centuries until the sixteenth and it exploited a stock the nature of which is obscure; some records survive and with the rise of the Hanseatic League the fishery must have been a large one for that period. The Dutch fishery rose as the Scanian faded, and it worked on the North Sea stocks from Shetland to the English Channel; the great period was in the late sixteenth and early seventeenth century. Some statistics survive from the mid eighteenth century. The fishery off Yarmouth was older and continued longer in record than any other; there are statistics of catches from the mid eighteenth century, but from 1826 to 1886 some essential records are missing, although sporadic statistics survive. The fourth preindustrial fishery in the North Sea was the Scottish; it existed in the Middle Ages, but its predominance dates from the early nineteenth century and statistics are complete.

Industrialization of the herring fisheries started in the mid nineteenth century, when nets were made by machinery and when cotton replaced hemp. The vessels increased in size and more nets were handled, but the most important innovation was the introduction of the steam drifter, first in Scotland. Herring were caught by trawlers from Fleetwood west of Scotland just before the First World War. Subsequent innovations were the introduction of large high headline trawls, pair trawls and light trawls between 1935 and 1955. Arbitrarily, I have taken the year 1930 as that by which catches were still taken predominantly by drift net. By the early 1950s catches were taken mainly by trawl and the ancient ancillary trades associated with drift-caught herring were starting to disappear.

The measures of weight or volume were based on the barrel (see p. ix) or on thousands of herring; a barrel was about 0.125 tonnes and there were about 6000 herring to a tonne. A last comprised 13200 herring, 2.46 tonnes; a Dutch last was about 1.5 tonnes and there were about 5000 fish to a tonne (J. J. Zijlstra, personal communication).

The Scanian fishery
(Mitchell 1864; Schäfer 1887; Lundberg 1886; Jenkins 1927)

The towns of Skanör, Falsterbø and Dragør in the southernmost province of Sweden, Scania, were the ports on which the Scanian fishery was based between the eleventh and sixteenth centuries. Hanseatic merchants from Lübeck, Stettin, Greifswald, Rostock and Danzig came to the Scanian ports to buy herring. The open boats were strongly built with crews of three to six men and they could carry up to twelve lasts. The fishery started in late August (perhaps early September today) off Malmö, and between September and November it lay of Falsterbø, which suggests a migration from the north; drift nets or set nets were used. Each merchant had his own area onshore and boats rode at anchor in the deep water alongside, waiting to unload. The fish were purchased on shore and then carried by wagon to the camps where women gutted them, washed them in sea water, roused them in salt and packed them in barrels, undamaged and uniformly layered of the same size. When the barrel was full it was allowed to stand for ten days, reopened and filled again. This was essentially the method until the drift net fishery in the North Sea died out in the 1960s.

There were peasants to haul and merchants to cure, and a huge crowd lived in wooden buildings along the shore. Each Hanseatic city had its own area marked out and in such *Vitten* (or camps) the herring were salted and packed; the *Vitte* of Kolberg was 480 yards (439 m) long by 160 yards (146 m) wide. Within each *Vitte*, each merchant had his own area. Herring were usually bought by the wagon; if a wagon could not take a full load of herring it was burned, the horses forfeited and a fine imposed. The first ship to anchor, loaded with herring, was the first to be allowed to depart. Vessels were loaded only in daylight; the fishermen paid a *baadsild* as a tax from 1376 to 1537; foreigners paid 10 Lübeck pfennigs per last as export duty, whereas the native Danes paid only 5 (Scania was then part of the kingdom of Denmark).

The Hanseatic League exchanged furs, salt, herring, flax, timber and tar from the northern cities for wool, leather and cloth from Venice. Salt springs were found in Scania, Pomerania and Rügen and herring fisheries developed in each place. In 1080, the fishery extended from the Sound to Falsterbø, and in the sixteenth century the herring left Scania 'on account of the people' (Jenkins 1927). The fishery was periodic and there were good yields in the second half of the thirteenth century and in the first thirty years of the fourteenth. The quantity of fish in the Sound was sometimes so great that on occasion 'the rudder could only be turned with difficulty and fish were taken by hand'. In 1368, 160 ships from Lübeck visited Scania and 34000 barrels (4250 tonnes) were brought back. In 1463, 20000 men worked off Scania with 501 boats and about 50000 tonnes were caught. In 1527, the sea in the Sound was filled with herring from bottom to surface and were not only caught with trawls but with wooden shovels at the surface. Quite large quantities of herring were caught off Scania for nearly 400 years, but the fishery may have reached its peak in the thirteenth and fourteenth centuries.

The nature of the stock is unknown. Höglund (1972) has shown that the fish which periodically visited the skerries on the Bohuslän coast of Sweden (north of Göteborg) in autumn and winter were North Sea autumn-spawning herring. Falsterbø and Skanör lie more than 320 km south of Göteborg. They may have been North Sea fish which entered the Sound in the Atlantic stream below the outgoing Baltic water in autumn – we do not know. There is no such fishery today and reasons for the collapse remain obscure. It collapsed when the Dutch fishery was rising, but the two events were probably independent, because the Dutch ports must have been further from the German market.

The East Anglian fishery

This account is taken from Mitchell (1864) and de Caux (1881), except where stated, and the latter part also from Nall (1866) and Jenkins (1927).

In AD 647, Felix I, bishop of the East Angles, was said to have built the church of St Benet on the Greenhill in Yarmouth 'to pray for the health and success of the fishermen that come to Yarmouth in the herring season' (Mitchell 1864). During the reign of Edward the Confessor (1042–1066), the town of Beccles, near Yarmouth, paid 30000 herring annually to the abbey in Bury St Edmunds (in West Suffolk). In 1067, the Free Fair for herring in Yarmouth was taken over by bailiffs appointed by the barons of the Cinque Ports (on the south-east coast of England). In 'Domesday Book', three salt pans were recorded at Gorleston near Yarmouth; de Montfort's manors in Suffolk yielded rents of herring. In 1087, Wilfred de Losinga, bishop of Norwich, built a chapel on Yarmouth sand and appointed a minister to pray for the success of the fishermen during the herring season. Swinden (1772), a Yarmouth historian, refers to the constant increase in the number of inhabitants of Yarmouth and to the great concourse of fishermen in 1259. In 1230, the abbott of St Albans bought great quantities of herring, which were stored in Yarmouth until resold to the advantage of the abbey. In 1242, Henry III bought herring from the men of Yarmouth. In 1277, the fishermen of that port were forbidden to dry the nets where the ships were being built. In 1286, a charter of land at Carlton (near Lowestoft) was paid for with twenty-four pasties of fresh herring at the start of the fishery each year. In 1290 the vessel which brought the Maid of Norway to Scotland was victualled in Yarmouth with dried herring. In 1295, Edward I gave permission to the Dutch to come to Yarmouth to fish on 28 September (9 October today). In 1302, by Royal edict, Yarmouth supplied 40 lasts (about 98 tonnes) to the king, 50 in 1303 and 40 in 1304. In 1305, 8 lasts and 1000 herring were sent from Yarmouth to the abbott and convent of Croyland in Lincoln-shire. In 1344, sixty foreign vessels (ten from Lombardy) entered Yarmouth to buy herring between 28 September (9 October today) and 30 October (10 November) at the Free Fair; up to 250 vessels worked from Yarmouth in that year.

Saul (1981) gives a good account of the Yarmouth fishery in the four-teenth century. The vessels displaced less than 30 tonnes, were single masted and open decked; they carried a crew of four to ten, were equipped with sails and oars and put to sea for one or two night's fishing, as did their successors in the twentieth century. They would catch 0.5–1 last per trip (about 2.5 tonnes) and 10–15 lasts in the whole season of ten to fourteen weeks. There were three categories of herring, white (gutted, salted and barrelled), dried (salted, sun dried and lightly smoked) and red (heavily smoked). From the murage account (a levy for the town wall) there were 500 vessels between August and November and 700 men from continental and English settlements. There were only forty Yarmouth boats. From the pledge accounts between 1331 and the early 1360s, there were 200–400 foreign vessels and forty English boats in Yarmouth each year. The numbers of Flemish vessels were 99 (1370), 61 (1373), 11 (1384), 1 (1390) and 26 (1398); only in two years after 1380 were there more than six vessels. From the customs records the annual catch between 1331 and 1348 amounted to 700–1800 lasts (1720–4428 tonnes) (2292 lasts in 1336, 403 in 1339, 1537 in 1348, 9 in 1349 and 1753 in 1358); only in one year after 1367 were more than 500 lasts recorded and only in one year after 1389 were more than 100 lasts recorded. This decline was associated with attacks by French and Spanish ships (Fulton 1911). The foreign merchants were Gascons, Spaniards, French, Florentines, and Germans (but there were no Hanseatic venturers). The home market was extensive, but from Yarmouth to Bardfield in Essex the cart took seven to fourteen days on their journey. This account shows how a professional historian can extract information from the records.

In 1357 the Statute of Herrings was enacted: no herring were to be bought or sold at sea, nor until 'the cable of the ship be fixed on shore' (Nall 1866); fishermen were to sell to whom they pleased from sunrise to setting; no hosteler was to pay more than 40 shillings per last; no fresh herring were to be bought by pykars (merchants) between St Michael (29 September) and St Martin (11 November); prices were set for fresh shotted (spawned) herring, fresh full (mature) herring, red shotted herring and red full herring. The dates of the fishery correspond well with those in the twentieth-century fishery. The statute was an emancipation from the market control by the Yarmouth burgesses to the advantage of the London merchants. In 1362, they sent 1 last of red herring, dried and well cleansed to the chapel of St George at Windsor. In 1373, John Botile of Lowestoft was fined for having bought 25 lasts from John Trample of Ostend. In 1428 herring were brought coastwise from Suffolk to Southampton (1539½ barrels) and, in 1430, 2590½ barrels.

In 1429, on Ash Wednesday, the Duke of Bedford sent 500 wagons of herring from Paris to Rouvray to the army under the Duke of Suffolk; they came under attack and Sir John Fastolf (Shakespeare's Falstaff) formed a laager and drove off the attackers in the Battle of the Herrings. Like monks and Lenten burghers, soldiers needed herring. In 1482, Edward IV invested Guardians to protect the fishermen on the coasts of Norfolk and Suffolk (Fulton 1911). In 1494, an Intercursus between Henry VII and the Duke of Burgundy (sovereign of the Low Countries) stated that fishermen of the two nations should fish freely everywhere, an anticipation of the Dutch lawyer, Grotius. Subsequently the English found it more profitable to buy from the Dutch.

In 1575, Manship (Palmer 1854) recorded that there were 600 sail in Yarmouth and that 'London, Kent, Essex, Suffolk, Norfolk, Cambridgeshire and Huntingdonshire were plentifully victualled'. There were fifty to sixty sail for trade with Italy, France, Spain, Flanders, Zealand,

Holland, Denmark, Norway and Russia. The herring were taken from 1 September (11 September today) to the last day of November (10 December today), 'swarming in sculls about the shore; there they were garbaged, salted, hanged and dried and by infinite numbers transported into the Levant and Mediterranean Seas where there be very good chaffer and right welcome merchandize'.

During the period of scientific investigation, it was established that the Downs stock of herring migrated south off Yarmouth to spawn near the Flemish banks and between the Somme and the Seine in northern France in November and December (Cushing and Bridger 1966), after which they migrated back into the North Sea in January and February. In 1030 certain salt works near Dieppe and Calais paid 5000 herring to the abbey of St Catherine near Rouen. In 1088, Robert Duke of Normandy gave permission to hold a fair one day during the herring fishery at Fécamp (near the spawning ground off Cap d'Antifer). Louis VII prohibited his subjects, in 1155, from buying anything at Estampes but mackerel and salted herring. In 1187, Philippe II decreed that Liège could deal in fresh and salted herring. The Comte de Blois gave 500 herring to the hospital of Beaugency in 1215; this trivial quantity was recorded, which indicates the value of herring in the long mediaeval winters. In 1383, the Annals of Dieppe record a fishery for herring off the coast between the Somme and the Seine particularly near St Valéry, not far from Cap d'Antifer, where the herring have spawned in the twentieth century. In 1468, in the peace treaty between Louis XI and the Duke of Burgundy, it was agreed that the French should not molest the fishery off Holland, Zeeland, Brabant, Flanders and Boulogne, the region for full and spent herring in the twentieth century.

In 1597 there were 220 fishing boats and about a thousand fishermen; the nets were spread on the Yarmouth Denes. But Raleigh (1603) noted that 30 000–40 000 lasts of herring were sold in Königsberg, Elbing, Stettin and Danzig, 10 000 sold in Denmark, Norway, Sweden and Lithuania, 1500 sold in Russia, 6000 sold to Staden, Hamburg, Bremen and Emden, 22 000 to Cleves, Juliers, Cologne and Frankfurt, 7000 to Liège, Zutphen, Deventer, Campen and Swell, 8090 to Guelderland, Artois, Hainault, Brabant, Flanders and Antwerp, and 5000 to Rouen and other parts of France. The total was 83 000 Dutch lasts or 125 000 tonnes. Not only is this an interesting view of European trade based on North Sea herring but Raleigh complained that British fishermen did not take part enough in the fishery. In 1611 and 1613 herring were so scarce in Yarmouth that the prices rose very high; in 1616 the export of herring in foreign bottoms was prohibited, but Yarmouth petitioned against it successfully (Samuel 1918).

Another advocate of increased English participation in the North Sea herring fishery was Tobias Gentleman, fisherman and mariner. In his pamphlet (Gentleman 1808), he contrasted the Dutch gear with 'the poor boats and sorry nets that our fishermen (use) in England'. He wrote that there were 600 great busses of 100 tonnes each and 400 of 50–60 tonnes. They probably landed up to 100 000 lasts (about 150 000 tonnes) and were protected by up to forty warships to 'waft and guard them'. There was a large number of small boats, of 20–50 tonnes, which 'go only for herrings in their season'. From the decayed towns of Kirkley and Layestof (Lowestoft) there were only six or seven boats. To Yarmouth came boats from the Cinque Ports, from Bridport and Lyme Regis, from Scarborough, Robin Hood's Bay and Durham. The western men fished for fresh herring and the northern men for red herring. Gentleman refers also to a French fleet based on Picardy and Normandy. By 1632, the Yarmouth fishery languished and there was no Free Fair. But in 1668, Sir Thomas Browne wrote that a 'Yarmouth man tells me . . . 90 vessels great and small went out this yeare to other parts with red herring' (Nall 1866).

For 500 years the Yarmouth fishery was international, based on the Free Fair (Nall 1866). It lasted through October and November, as in the twentieth century; full and spent herring were caught there and they were cured as salted or red herring. They were used as 'lenten stuffe' (as Thomas Nashe (1599; see McKerrow 1901–10), the Lowestoft poet called it) for winter fast days and the long days of Lent. They were used as rent and were distributed widely, as for example, to the abbeys where records were kept.

Defoe (1724–6) described the Fair and saw 110 barks and fishing vessels on one tide, 'all loaded with herring'. The fish were brought ashore 'by open boats, which they call cobles and which often bring in two or three lasts at a time'. The barks often 'bring in ten lasts apiece . . . Some have said that the towns of Yarmouth and Leostoft only have taken 40,000 lasts in a season; I will not venture to confirm that report.' The merchants cured 40 000 barrels (5500 tonnes) of merchantable red herrings in one season, 'over and above all the herrings consumed in the country towns for thirty miles from the sea, whither very great quantities are carried every tide during the whole season'.

In 1751 there were 250 Dutch busses (of 80 tonnes each) which landed 5000 lasts during the season, 120 Scheveningen boats (of 30 tonnes each) which landed 900 lasts and 120 French boats (of 100 tonnes each) which

landed 3000 lasts, nearly 22000 tonnes in all (Nall 1866), to satisfy the
English market. For the second half of the eighteenth century, much
information is given in the *Reports of the Select Committee on Fisheries* in
1785, 1786, 1798 and 1800. John Shelley (quoted by Nall (1866)) said that
each fishing vessel shot 90–100 nets, each 20 yards (18 m) long and 6 yards
(5.5 m) deep, fastened to six warps of 240 m each (Figure 34). The fishing
grounds lay from 50 km to the north of Yarmouth to south of the Foreland, on the Banks of Flanders and the fishery lasted from 21 September
to 25 November. Gear, grounds and seasons were those common nearly
200 years later. In 1760, there were 205 vessels, 94 in 1783 and 100 in 1785.
Shelley said that the cause of decline was the great loss of fish in 1760 and
the great advance in the price of cordage and netting. The 1785 Report
gives statistics of barrels of red herring, both exported and for the home
market, between 1739 and 1782; barrels of white herring were reported
from 1751 to 1782. Gillingwater (1790) recorded catches in barrels at
Lowestoft between 1748 and 1789, including catches made by northern
(from Scarborough) and western (from the Cinque Ports and Bridport)
fishers. The catches are recorded (in tonnes) in Figure 35. Cushing (1968)
analysed Gillingwater's material in detail and was able to indicate which
were the notable year classes and how the American War of Independence affected the fishery.

In 1785, the Dutch boats came to the Free Fair a few days before
21 September and the Sunday before the Fair opened was called Dutch
Sunday. Their *schuyts* were small-decked, flat-bottomed, beamy vessels
adorned with paintings; they had yellow sails and striped pennants

Figure 34. French drifters working for herring, possibly off Boulogne (Duhamel
du Monceau 1769).

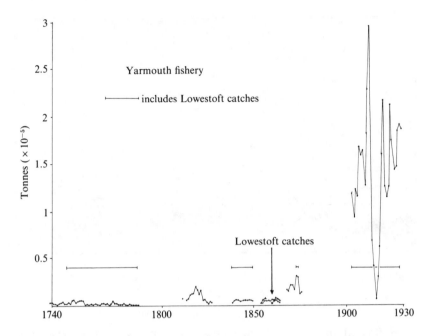

Figure 35. The East Anglian fishery from 1740 to 1928; catches at Lowestoft are taken from Gillingwater (1790, see Cushing 1968), Nall (1866) and the *Bulletins Statistiques du Conseil International pour l'Exploration de la Mer* for 1903–28.

(Palmer 1854). In 1798, Samuel Hobbins told the Parliamentary Commission that there were only eighty boats and 'since the loss of the Italian market, Yarmouth lost two thirds of its fleet, Lowestoft a quarter and the Dutch fleet (in Yarmouth) was reduced to twenty sail'. Such was the effect of the Napoleonic wars.

The vessels in the East Anglian fishery in 1800 were large, bluff bowed and three masted, with a large lug sail on each. They were derived from the English herring buss of 1677 (which did not resemble the Dutch; see Figure 36). By 1849 a three-lugsail lugger had been developed with a rounded counter. On the north-east coast there was also the five man boat, 18 m in length, which carried two cobles (small boats) on deck. In 1874, the lugger's main mast was discarded and the smack functioned as a drifter with the foremast lowered for fishing (White 1950). The crew comprised a master, mate, hawseman, whaleman, net ropeman, net stowerman and five or six capstan men, and they were engaged for a whole season. The vessels fished for up to 50–60 km offshore. Between 80 and 100 nets were shot, each 21 yards (19 m) long and 8½ yards (8 m)

Figure 36. The English buss or Dutch smack (Duhamel du Monceau 1769).

deep, floated up by corks from a messenger rope by 'seizings', each 8 m in length. When the vessels were victualled and loaded with salt they could stay at sea for a week. When landed, the fish were salted for two or three days, washed in large towers filled with fresh water, split, and smoked in smoking towers. In the first decades of the nineteenth century there were about 200 vessels and 2000 fishermen supported by 400 braiders, beetsters, towers, reivers, ferrymen, carpenters, caulkers and spinners (Chambers 1829).

In 1808 the Board of the British White Herring Fishery was established. Their statistics from 1810 to 1875 were very complete for Scottish ports and for some ports in the north of England and the Isle of Man. Unfortunately the returns from other English ports gave figures for white herring only (the salted herring); the fresh and the red herring were excluded. The Scottish fishery was predominantly for white herring and the Board was named for the white herring. By 1849, the English statistics were dropped completely and none can be traced as a continuous series until 1886.

Palmer (1854) wrote that 7000–8000 lasts (19 680 tonnes) were taken by 193 vessels in 1817 and that, up to 1854, an average catch of 3500 lasts (8610 tonnes) was a reasonable estimate and that about 30 000 -40 000 barrels (3750–5000 tonnes) were exported each year. The season of 1808 was very productive and one vessel landed 62 lasts in the season. In 1809, 101 vessels caught 3500 lasts. By 1853 the drifter *Perseverance* brought in

126 lasts in one season. Probably 24 600 tonnes were landed, 15 000 barrels (1875 tonnes) exported and as much as 12 000 tonnes were distributed by railway (which had just reached Yarmouth). Nall (1866) wrote: 'before the Norfolk railway was constructed the conveyance of fish from Yarmouth to London was entirely conducted in light vans with post horses and was represented by a bulk of 2000 tons a year. At present, 2000 tons can be carried in a fortnight.'

Nall gives catches at Lowestoft for certain years between 1839 and 1862; Holdsworth (1874) recorded catches at Yarmouth from 1868 to 1876 and at Lowestoft in 1861–4 and in 1873–4. In 1853, 12 000 tonnes were sent into the country by rail from London and many thousands of barrels were sent coastwise by steamers. In 1854, 20 000 tonnes were despatched from Yarmouth by rail alone. The statistics between 1739 and 1886 are shown in Figure 35; it is not a proper collection, as was available after 1886. In the eighteenth century, 5000–10 000 tonnes were landed, 10 000–30 000 tonnes in the first half of the nineteenth century and more than 50 000 tonnes in the 1870s.

The Scotch nets (machine made of cotton) were probably in use off East Anglia in mid century, made in Musselburgh (in Scotland) or in Bridport (in the West Country). Scottish vessels first came to the East Anglian fishery in about 1860. In 1882 the *Rob Roy*, a steam liner, was built in Leith; subsequent steam vessels were used for lining and drifting and she may have been the first steam drifter (Wilson 1965). In 1884, the steam capstan was invented by Elliott and Garrood of Beccles, near Yarmouth; the steam was fed into the capstan through its hollow spindle and so the warp was free. In 1897, the *Newark Castle* was launched on the north-east coast of England (Wilson 1965) and the *Consolation* was built in Lowestoft, the first English steam drifters (Goodey 1976). The nets were now 50 m long (mounted on the headrope at 32 m) and 13 m deep; up to ninety or 100 nets were shot, making a fleet of about 2.5 km in length. The big French drifters were shooting up to 280 nets at this time. These were the major steps in industrialization, the machine-made cotton nets, the steam capstan and lastly the steam drifter itself. The rise of the steam drifter between 1897 and 1930, together with the decline in sailing drifters in England and Wales (which includes vessels outside East Anglia), is shown in Figure 41. They reached a peak in 1913 and declined after the First World War. They were replaced by motor vessels.

The East Anglian fishery yielded its greatest catch in 1913 when 266 560 tonnes were landed in the ports of Yarmouth and Lowestoft by nearly 1600 vessels. Most of the catch was exported in barrels to Germany, Russia and to the Levant. After the First World War, the Russian market was lost and there was competition from German trawlers by the mid 1930s. After the Second World War, catches were reduced in competition with trawlers for fish meal and by the 1960s the drifters were being laid up; the last drifter sailed from Yarmouth in 1967.

The East Anglian drift net fishery lasted from the eleventh century (and perhaps earlier) to the twentieth. It was obviously profitable in the Middle Ages, but the quantities landed were not very great, perhaps 5000 tonnes in the fourteenth century; however, with poor transport and a low population the quantity was as much as could be handled. It is likely that throughout their period of supremacy the Dutch landed the fish for the English market. The fishery was poor in the early seventeenth century, but from Defoe's (1724–6) account, in the first half of the eighteenth century, it was profitable. Industrialization from the mid nineteenth century generated great increases, and the period of the steam drifter was that of greatest prosperity.

The Dutch fishery (Beaujon 1884; Jenkins 1927)

In 1163, fishermen from Kampen and Harderwijk (on the Zuider Zee) went to Scania. At about the same time fishing for herring started in the Meuse and at Brielle (Mitchell 1864). In 1295, Edward I of England said that 'many people from Holland, Zealand and Friesland, who are our friends, will shortly come and fish in our sea off Yarmouth' (Mitchell 1864). In 1368, men from Amsterdam, Enkhuisen and Wieringen were allowed to fish off Scania. There were markets for herring and other fish in Brouwershaven (1344), Naarden (1355) and Katwijk (1388). Willem Beukelsen of Bievliet (who died in 1347, 1397 or 1401) devised the method of opening herring with a single cut; fish were taken straight from the sea, salted and packed in barrels at sea. Such an operation could be sustained only in a vessel such as a buss, which was keeled and decked. In 1416, the first large herring drift net was made in Holland. In 1424, the first regulations were published on curing herring and marking barrels. In 1429, the Scots were selling herring to the Dutch (Mitchell 1864). In 1519, the edict of 18 May of Charles V was published: the branding of barrels was obligatory, marked by the cooper and separately by the skipper. Only new barrels were used and the brands were controlled by the *Keuremeisters*. Only salt made at the mouths of Dutch rivers (moor salt) was to be used. Herring caught before St James' Day (25 July, or 4 August today) were inscribed in separate barrels with St James' shell (in 1536 this date was changed to 24 June, or 4 July today). There were rules on packing herring and they had to be pickled (moistened with sea water) every fortnight. The insistence on quality was an important factor in the development of the Dutch fishery.

In 1556, the main herring ports were Delft, Rotterdam, Schiedam, Brielle, Enkhuizen, Wormer, Jisp and Grootebroek. The deputies of the Herring Fishery agreed in 1558 to have every tenth buss equipped for war and, in 1575, twelve convoy vessels were chartered. This tax was taken from each last (1.5 tonnes) landed. In 1578 and 1580, rules for finding lost fleets of nets were published together with the signals to be hoisted when the nets were shot. In 1560, Guiccardini (quoted by Samuel 1918) recorded that 700 vessels from Friesland, Holland, Zealand and Flanders made three voyages each year and took 70 lasts per vessel per year (Jenkins 1927). Thus, towards the end of the sixteenth century, the Dutch fleet may have caught as much as 75 000 tonnes.

During the period of the Dutch republic, the fishery took place in the second half of the year off the coasts of England and Scotland. The busses had a crew of fourteen to sixteen, and the herring were gutted, cured and salted on board; the keel was about 16 m in length. The drift nets were 64 m in length and made of hemp, and a fleet comprised about forty nets. In 1621, the busses were fitted out at the beginning of May. On St John's Day (24 June, or 4 July today) 600 vessels of 100–120 tonnes each started fishing near Shetland, Fair Isle and Buchan Ness and continued until St James' Day (see above). Until Elevation Day (14 September, or 24 September today), they fished the deep water off Yarmouth (Jenkins 1927). In 1604, the *Ventjagers*, or fast sailing vessels, brought the New Herring back to Holland between St John's and St James' Days. The number of busses in the first decade of the century was quoted as 1000 (de la Court, quoted by Beaujon (1884)), 3000 (Raleigh 1603), 1500 in 1610 and 2000 in 1620 (den Koopman; quoted by Beaujon (1884); Beaujon (1884) wrote that there probably were not more than 2000. At this time the busses took 40 lasts per season each, perhaps 120 000 tonnes each year.

During this period there was conflict with the privateers from Dunkirk; in 1625, 100 busses from Enkheizen fell to them. In 1633, there were 1500 busses off Unst in Shetland, protected by twenty waffers, each armed with twenty guns (Mitchell 1864). James I had declared the *Mare Clausum* in 1609, and in 1622 a British admiral collected taxes from Dutch fishermen on the high seas. The Navigation Act (1651) forbade any import of fish save in English vessels, which reduced the Dutch carrying trade considerably. War broke out between England and Holland in 1652–4 and in 1653 2000 busses were kept in port for safety. However, during the reign of Charles II the Dutch brought prosperity to Yarmouth with a 1000 sail and 10000 people ashore (Jenkins 1927). There was war again between 1672 and 1677, and the Dunkirk privateers survived until 1678; there were conflicts with the English until the peace of Breda in 1667. It is not surprising that there were complaints on the quality of herring from Stettin and Danzig in 1683 and 1687 (Beaujon 1884).

In 1702 the War of the Spanish Succession started, and in 1703 four French men o' war destroyed 400 busses and their armed convoys in Bressay Sound. Between 1652 and the Treaty of Utrecht in 1713, the fleet of busses must have sustained continuing losses at sea. In 1715, the Dutch complained about Scottish herring in Hamburg. By 1731, their brands had disappeared from that port, where the market was open to the Scots without restriction; indeed in 1721 the magistrates in Stettin remonstrated with the Dutch about the quality of the branded herring. The trade was faring ill by 1750, 'most busses sailed money overboard, some returned neither gain nor loss and a very few brought a small clear profit' (Beaujon 1884). However, in 1751 there were 400 busses and 120 French vessels within sight of Yarmouth (Jenkins 1927).

The development of the fishery from 1750 to 1930 is shown in Figures 37 and 38. Figure 37 gives the number of busses from 1750 to 1851. The fleet of busses was reduced during the Seven Years War (1756–63) and during the latter part of the American War of Independence; between 1783 and 1794 the number recovered to about 200. After the Napoleonic wars the number of busses was reduced from about 160 to about 100 in the early 1850s. After the wars, the numbers of *bomschuyten* (flat-bottomed boats which can be hauled up on the beach) increased sharply in the 1840s. From 1856 onward, the number of keeled vessels (busses were being replaced by luggers) remained steady until they increased in the 1870s. The *bomschuyten* increased in numbers at the same time but later were replaced by the luggers. From 1814 the catches remained relatively low, between 5000 and 10000 tonnes each year until the 1860s and 1870s. Figure 38(*a*) shows the very sharp increase to about 70000 tonnes, sustained and augmented until the 1930s. Figure 38(*b*) gives the catches per unit of effort between 1814 and 1908. There is a general increase by a factor of about 4 for the keeled vessels and of about 5 for the *bomschuyten*. Cotton nets were introduced to the *bomschuyten* in 1857 and by 1865 one-third of the keeled boats had them. In 1866 the first lugger, based on a Boulogne model, was introduced. Beaujon (1884) showed that the hookers and sloops were replaced by luggers. The cotton nets were lighter and much more efficient than the old hemp ones and they were larger; further, as the vessels became somewhat larger, more nets were shot. The increment in catch per unit of effort was probably one of efficiency. A

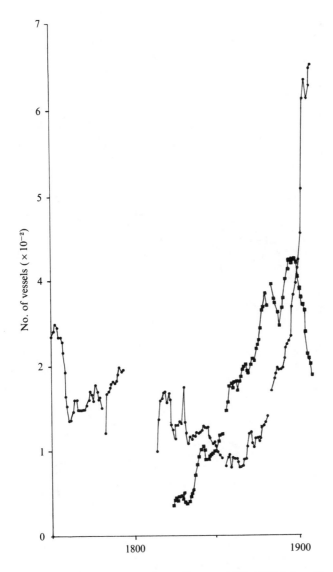

Figure 37. The Dutch fishery; number of busses, 1750–1853 (●) and keeled boats 1856–1908 (■); number of *bomschuyten* (●) (Beaujon 1884).

natural change might have occurred but perhaps the change in efficiency was sufficient.

In 1818 the Dutch regulations had been simplified, if their monopolistic intention remained: no foreigner to share in the herring fishery, no Dutchman to share in foreign herring fisheries and no foreign currency to

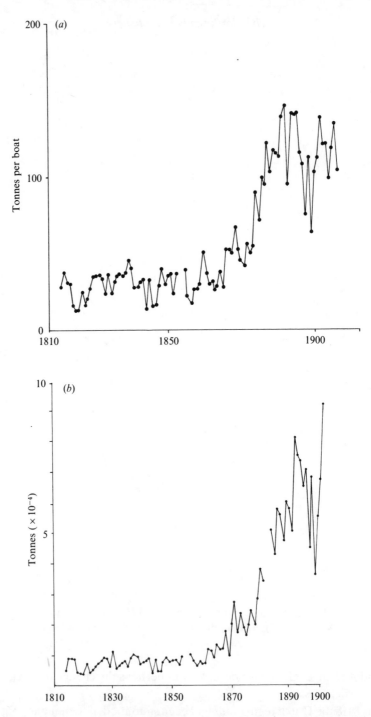

Figure 38. (*a*) Stock densities in tonnes per boat in the Dutch fishery from 1814 to 1908; (*b*) total catch in lasts, 1814–53; number of barrels, 1856–1881; number of lasts landed, 1882–1902 (Beaujon, 1884; *Verslag van den Staat Nederlandsches Zeevisscherlijen* for 1885–1908).

enter Holland. In 1822, four new brands were introduced: *fulls*, *maatjes* (mature), *ijlen* (lean or spent) and *kurtzoek* (light cure). But the three monopolies – fishing and curing, carrying and selling – were abolished in 1858 (bounties had been abolished in 1850). The revival started with cotton nets and new vessels.

The Dutch fishery started with the buss, the long net and the quality control of curing at sea. The fleet developed in the fifteenth and sixteenth centuries, reached its peak of 2000 vessels in the first half of the seventeenth century and may have landed as much as 120000 tonnes each year. Wars and privateers in the second half of that century reduced the fleet to 200 busses in 1793. In the early and middle nineteenth century, catches were much lower, but with the new methods they increased to nearly 100000 tonnes in the first decade of the twentieth century (Figure 38(*b*)).

The Scottish fishery
(Mitchell 1864; de Caux 1881; Samuel 1918; Jenkins 1927)

In 1138, David I of Scotland granted to the abbey of Holyrood in Edinburgh the right to fish herring off Renfrew, and in 1153 the right was established for it to sell salt and herring at sea. At this time, Scottish and Belgian fishermen worked off the Island of May in the Firth of Forth. In 1415, the Scots put forward the exclusive right to fish off their coasts. In 1429, herring were plentiful off Inverness, in Loch Fyne and off the mouths of the Dee, the Tay and the Tweed. In 1471, James III encouraged the building of Dutch busses: 'France, Flanders, Zealand, Holland and mekill of Almany comis with sundry flotis, passand in the time of Lentrown throu the seis Mediterrane aye selland their fische to thair grit profit'. Also, 'in Loch fine is mair plenti of herring than is in ony seis of Albion' (Mitchell 1864). In 1488, James IV decreed that 'strangers buy na fish but salted and barrelled'. The Scots interrupted the Dutch fishery in 1532, and Robert Fogo of Leith took many busses with warships. In 1540, an Act was passed which limited the manner of sale so that the home market was supplied first and barrels were to be marked by the cooper.

In 1584, it was decreed that herring caught within the Firth of Forth be sent to Leith, those caught between Fife and the mouth of the Dee should be sent to Dundee and Perth, and those caught west of Dumbarton should be distributed to the free burgesses. In 1587, a fishery was recorded in the northern Highlands and in 1600 an act was passed prohibiting the export of herring before Old Michaelmas (11 October, or 21 October today). In 1609, James VI (James I of Great Britain) prohibited foreigners from fishing on the coasts of Britain, the *Mare Clausum*. By 1661, Charles II renewed the prohibition and the Dutch paid £10000 as a licence. In 1667 Sir Robert Sibbald reported that 600 boats worked in the Clyde and 168 off the coast of Fife. In 1693, all herring exported were to be packed in well-seasoned barrels of maple or oak; each barrel should contain 8 gallons and 2 pints Scots (about 37.5 l), be free of all white wood and worm holes, be properly examined and branded, cured with French, Bay or Spanish salt, and the herring inspected and exported to the Sound.

Thus, Scottish herring were salted, redded and barrelled as in other early fisheries and the quality controls resemble those of the Dutch, but the fish were cured ashore. The herring were caught in the Firth of Forth, in the Clyde, in Loch Fyne and in the northern Highlands, off the Dee, the Tay and the Tweed; all are sites of fisheries as we have known them in recent history. There is also an indication of the modern rights of the coastal state with a *Mare Clausum* out to 200 miles, perhaps because the Scots saw Dutch busses offshore, as people in developing countries saw large trawlers in their waters in recent years before Exclusive Economic Zones were declared in the early months of 1977.

In the second half of the eighteenth century a number of attempts were made to encourage fisheries off the Scottish coasts and off Yarmouth (Dunlop 1978). In 1750, the Society for the Free British Fishery equipped busses for the Shetland and Yarmouth fisheries and bounties were provided. The venture was not very successful although an export trade was sustained; Adam Smith (in *The Wealth of Nations*; see Jenkins 1927) remarked that the fishermen fished for bounties but Thomson (1849) pointed out that this was because the bounty was based on the size of vessel and not on the size of the catch. In 1786, the British Fisheries Society developed harbours on the west coast, but their most important venture was to build the harbour in Wick, completed in 1811 (it was destroyed in 1870). For much of the nineteenth century, large quantities of herring were passed through that harbour. In 1784, the Dutch cure still held the European market and the international fleet off the Scottish coasts comprised 166 Dutch, 44 Prussian, 29 Danish, 24 Flemish and 7 French busses (Dunlop 1978).

Anderson (1785) reported on the *Present state of the Hebrides and western coasts of Scotland*. In 1773, 'herring swam so thick in Loch Torridon that the boats of about 250 "busses" . . . together with an immense number of country boats . . . were often twice loaded in a night . . . [This] went on for two months'. He gives similar accounts for Loch Slapan and Loch Earn. Anderson's busses carried two or three boats which were large cobles, 9–10 m in length; their nets were of two-ply twine (hempen), 15–18 m long and 5.5 m deep.

In the last quarter of the eighteenth century on the east coast herring were caught only on the south coast of the Firth of Forth. The fishery off Caithness developed as Forth merchants went north to buy cod and they found herring. By 1789, 100 boats worked out of Wick and 10000 barrels (*c*. 1250 tonnes) were packed. In 1800, most boats were 5 m long in the keel and each shot eight or ten nets, each 24 m in length. By the 1820s there were more than 1000 boats at Wick and the harbour became over-crowded (Gray 1978). The records in the 1796 Parliamentary Report can be separated into those from the North Sea and from the west coast; the North Sea catches from 1771 to 1797 are shown in Figure 39.

In 1794 many vessels came to the Firth of Forth from many of the North Sea ports of Scotland; in October and November, 'shoals formed an oblong square, from Burntisland to several miles westward on the North side of the Firth . . . [There was] abundant fishing for some months until the middle of March . . . [Catches were] sent by fast sailing vessels to London' (Mitchell 1864). There were twenty-two smoke houses in Burntisland for smoking red herring. This fishery became much less important after 1805. The most important action was Act 48 of George III

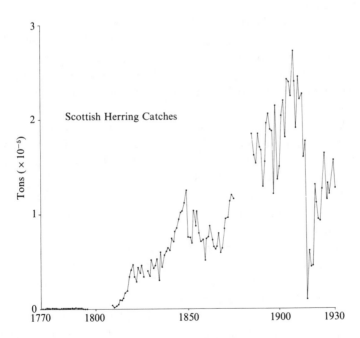

Figure 39. Catches in the Scottish fishery, 1771–96 (*Parliamentary Reports*, 1796); 1809–76 (*Reports of the British White Herring Board*); 1755–1902 (*Annual Reports of the Fishery Board for Scotland*); 1903–30 (*Bulletins Statistiques du Conseil International pour l'Exploration de la Mer*). (1 ton = 1.016 tonnes.)

in 1808, which provided bounties for fishing vessels. There is little doubt that such bounties were effective in starting the Scottish fisheries, despite Adam Smith's comments (see above).

To qualify for a bounty, the larger vessels were to rendezvous at Bressay shoal off Shetland before 22 June, where they were inspected. No nets were to be shot before 26 June and the fishery stopped on 15 September. The nets were to be shot and hauled without the use of a small boat. The herring were cured on board. The master had to keep a journal which was verified on oath before he was entitled to a bounty. The Admiralty appointed a naval officer to superintend the fishery.

In the early nineteenth century, the Scots fished from open boats and the blankets for the fishermen were also used as sails. There were four men in each boat and the fleet of nets was stored in four barrels, one for each fisherman. Each net was 11 m long and 3.7 m deep. The nets were made at home; the women spun the twine from hemp and the fishermen made the nets. In 1821, half-decked ships were introduced, and in 1839 there were 200 half-deckers in addition to the traditional sixareens (for six men) at Shetland (Coull 1983). By 1830, keels had lengthened to 9 m so that a longer fleet could be shot (Gray 1978). They caught on average 0.5 to 1 tonne each night.

Net-making machinery had been introduced in the 1820s (Gray 1978). In 1840 Paterson of Musselburgh introduced cotton nets knotted with the fisherman's knot (devised by William Ritchie); the fleet of nets comprised twenty, each 91 m in length and 7.5 m deep. The cotton nets were light and they were floated up from a heavy messenger rope attached to the vessel (March 1952).

Buckland, Walpole and Young (1878) in their report on Scottish fisheries summarized much of the statistics given in the Board's reports. In 1861 the curers of Montrose encouraged the fishermen to go further to sea 30–50 km, instead of 11, where the fish were finer. The ports of Frazerburgh, Peterhead and Aberdeen increased their catches, whereas those of the smaller ports around the Moray Firth, such as Lybster, Buckie and Helmsdale, declined. This was probably because the exporting vessels entered the larger ports more readily. In the Firth of Forth, catches declined from about 5600 tonnes in 1867 to 750 tonnes in 1876. In the long term, the decline at Wick was the most dramatic, because until mid century it had been the dominant port.

After the Napoleonic wars, the continental trade became important and gradually replaced the earlier exports to the West Indies and to Ireland. By 1850, this trade predominated and replaced the Dutch cure all over eastern Europe, particularly in Stettin. The Dutch cure remained the best in quality but was more expensive, and the cheaper Scottish cure supplied the major markets. The bounties (2 shillings per barrel and 2 shillings and 4 pence for each barrel exported in 1786) stimulated and sustained the fishery until 1829 when they were abandoned and the fishery supported itself on its markets (Gray 1978).

One of the interesting points put forward by Buckland *et al.* (1878) was the demonstration that the herring stocks could not be overfished. Cod often have five or six herring in their guts; 2 herrings × 210 days, 420 herring each year. Three and a half million cod and ling are caught each year, perhaps one twentieth of the stock: $70.10^6 \times 420 = 29.4 \times 10^9$ or twelve times the catch of 24×10^8. Also, there are 10 000 gannets on Ailsa Craig in the Clyde each of which eat six herring each day; as there are fifty times as many gannets elsewhere (now known to be a gross overestimate), 1.11×10^9 are eaten. These arguments were presented to counter the fears of the fishermen that they were overfishing the herring stocks. Burd (1978) shows that in the 1920s $F = 0.3$ and $F/Z = 0.75$ (where F is the instantaneous coefficient of fishing and Z that of total mortality), when catches amounted to 0.5 million tonnes. In Buckland's time, catches were not more than 0.1 million tonnes and his conclusion was probably right, particularly if the present estimate of natural mortality (M) of 0.1 is much lower than when the gadoid stocks were lightly exploited.

In the 1860s, the first boats from Fife went to the autumn fishery off East Anglia. In 1890, they started to come from other ports. The 'Fifie' was a longer vessel (13–17 m on the keel) with a lugsail and the 'Zulu' (1879, 18 m on the keel) was a very fast vessel (March 1952). The *Rob Roy* was the first steam drifter/liner built in 1882 in Leith. The rise of the steam drifter and the decline of the sailing drifter is shown in Figure 40. This major step in industrialization occurred in the first decade of the twentieth century. After the Second World War the numbers of steam drifters started to decline, but in Scotland in particular they were replaced and augmented by motor drifters. The point made in Figure 40 is merely that the traditional sailing boat, if much larger than thirty years before, was replaced by the steam drifter.

In the first decade of the twentieth century the catches surpassed 200 000 tonnes, but after the First World War they were reduced, with the loss of markets, particularly in the USSR and eastern Europe.

Herring were caught off Scottish coasts many centuries ago particularly in the Forth, the Clyde and in the sea lochs on the west coast. The North Sea fishery from the east coast ports north of the Tay did not start until 1771. Stimulated by bounties, the Scottish fishermen took on the German

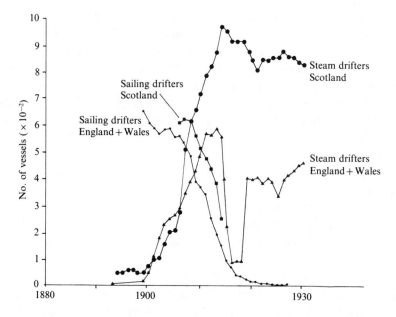

Figure 40. The rise and fall of the steam drifter; number of sailing drifters in England and Wales (●) and in Scotland (■); number of steam drifters in Scotland (●) and in England and Wales ▲). The records are taken from the *Sea Fisheries Statistical Tables of the Ministry of Agriculture, Fisheries and Food* and from the *Annual Reports of the Fishery Board for Scotland*.

and Russian markets at a time when a general industrialization opened up transport and markets. Industrialization of the fishery in Scotland meant the introduction of machine-made cotton nets, the lengthening of the fleets of nets, the increase in size of vessel and finally the development of the steam drifter.

The North Sea

The herring fishery in the North Sea has always been international, off Scania, off Yarmouth or on the high seas. The list of nationalities present at the Free Fair, in Scania or with the Dutch fleet offshore, shows that fishermen from many countries took part, but the catches from a single port are probably low. The numbers of people were also much fewer, much less than a tenth of the present population in the North Sea coastal states.

The Scanian fishery supported the Hanseatic League for perhaps 400 years and catches of up to 50 000 tonnes were landed in a year; the Hansa

Table 5. *Roughly estimated annual catches from the North Sea*
(in 10³ tonnes)

Century	Scanian	East Anglian	Dutch	Scottish
12				
13	Up to 50			
14		5–10		
15				
16			125?	
17		Low in early century	120	
18		60		
19		10–50	100	200
20		750		

merchants worked right across Europe, living on herring as they travelled. The Free Fair at Yarmouth hosted fishermen from all around the southern North Sea and eastern English Channel and merchants came there from all over Europe. The foreign fishermen worked the English market as well as their own. The Dutch fishery operated fully offshore, curing the fish at sea. At its peak in the seventeenth century, annual catches of up to 120000 tonnes may have been taken. But its effort was reduced by a succession of wars, particularly in the eighteenth century. In the nineteenth century, the Scottish fishery reached the same level but it was fed by the processes of industrialization. Table 5 gives rough estimates of annual North Sea catches. The replacement of hemp nets by machine-made cotton ones and the considerable increase in size of vessel meant that a greater number of more efficient nets were shot, yielding greater catches. Such an increment in catch per unit of effort was shown in the Dutch fishery as the cotton nets were introduced. The steam drifter with its steam-driven capstan was another industrial innovation; a long fleet of nets could be hauled more quickly and the vessel could reach port independently of the wind. Hence, its seasonal catching power was increased considerably. In the final stage, Scottish, English, Dutch, French and German steam drifters took half a million tonnes of herring each year from the North Sea between 1890 and 1914 (Figure 41). An inset (Figure 41, lower) shows the Swedish catches on the Bohuslän coast; perhaps part of the increment in the first decade of the century in the North Sea is the decrement off Sweden.

Before the First World War off the west coast of Scotland herring were taken by large trawlers. During the interwar years, the Germans

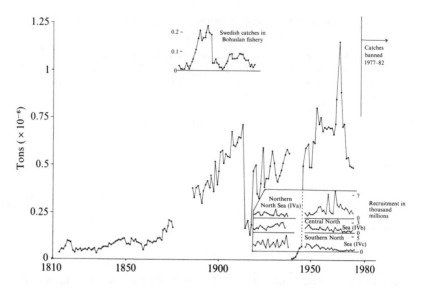

Figure 41. Catches from the North Sea, 1814–1977, including those from Figures 35, 38 and 39; in addition records from Schnakenbeck (1953) are included. Inset (a) shows the Swedish catches of North Sea herring between 1876 and 1919 (Devold 1963). Inset (b) shows the recruitment to the three areas in the North Sea, IVa, IVb and IVc estimated by sequential analysis in thousand millions (Burd 1978). (1 ton = 1.016 tonne.)

developed high headline trawls for catching herring, and by the end of the 1930s large trawlers were working on the Fladen ground and on the spawning grounds around the Dogger Bank. After the Second World War, the method was adopted by French, Belgian, Dutch and Polish fishermen and, with the Germans, they worked the spawning grounds in the central and southern North Sea. By the mid fifties, the trawl catches of adult fish for fish meal were much greater than the catches by drift net (Cushing and Bridger 1966). On the Sandettié spawning ground (between the Sandettié and Ruytingen Banks about 24 km from Dover) in December 1957, I saw about 150 large trawlers working in six parallel lines each about 180 m apart. In the same decade, Dutch and Belgian fishermen developed the pair trawl for catching herring in midwater on or near the spawning grounds in the southern North Sea.

In the eastern North Sea, the immature North Sea herring live on the Blöden ground (east of the Dogger Bank) and in the Skagerak. Danish fishermen developed light trawls for catching the little fish for fish meal. By 1955, about 100000 tonnes of immature herring were caught each year. This was the start of the Danish industrial fishery for young herring,

Norway pout, sandeels and sprats which, by 1974, landed nearly two million tonnes of fish for fish meal. Because a fair proportion of the immature herring had not survived their first summer, the numbers caught were very large, particularly in the Skagerak.

There are three or four spawning grounds of autumn spawning herring in the North Sea. During the 1950s, the southern group, the Downs stock, was reduced by fishing on the adults to a single year class and at the same time recruitment (the year class (see p. 213) which joins the adult stock) was being reduced (see Figure 41, lower inset). By 1963, this stock was obviously suffering from recruitment overfishing by reduction of the adult stock and by capture of potential recruits on the nursery ground; the number of recruits can be reduced by heavy fishing of the adults. By the early 1970s, the other North Sea stocks had also been reduced to recruitment overfishing. But between 1950 and 1969, the Buchan stock (in the northern North Sea) benefited from natural increases in recruitment (see Figure 41, upper inset; Burd 1978); this event may have been associated with the increase of *Calanus finmarchicus* in the eastern North Sea from 1950 onwards (Cushing 1982). In 1977, catches of North Sea herring were banned, and by 1982 stocks in the central and southern North Sea (Dogger and Downs) had recovered, which confirmed the diagnosis of recruitment overfishing.

The traditional markets for herring lay all over Europe and parts of the Mediterranean; pickled herring went to the northern countries and red herring to Italy and the Levant. Such markets expanded with the more general industrialization of the mid nineteenth century. The industrialization of the fishery started with the mechanization of net making and the introduction of larger boats, culminating in the steam drifter and later the motor drifter. Between the First and Second World Wars, the Russian market was lost and the German market was transferred from British driftermen to German trawlermen. After 1945, a large new market opened up, that for fish meal for animal feeding stuffs. The trawl fisheries on the spawning grounds, on the nursery grounds and in the midwater supplied this avid market. The later industrialization of the herring fishery in the North Sea followed the industrialization of agriculture with its dependence upon fish meal as an essential part of animal food.

The drift net gave way to the trawl. With it vanished the ancillary trades, netmakers, rope makers, sail makers, Scots fisher girls, people who smoked kippers, bloaters and red herring and so on. The oaken skeletons of drifters rotted in the creeks and the old driftermen talked to the tape recorders of the oral historians. The long history that started in Skanör and Falsterbø ended 800 years later in food for animals.

5

The first industrialization of fisheries

The early industries used both energy and labour freely. A textile factory or a railway employed many people to work the machines driven by steam. Output in quantity of yarn or in distance travelled increased dramatically in the early days of the industrial revolution. The machines spread quickly where there was labour available. The first steam boat, the *Charlotte Dundas*, was seen on the Clyde in 1802 and another, the *Savannah*, crossed the Atlantic in 1838. The industrial revolution affected the fisheries in Britain from the period just after the Napoleonic wars by providing more extensive markets. The method of capture, however, remained unmodified for another sixty years or so, for fishing boats were not driven by steam until the last two or three decades of the nineteenth century. The mechanization of fishing and the standardization of fisheries products came later in the industrial revolution and the reason is an obvious one, that, however abundant, coal costs money but wind and tide are free.

When the fisheries were finally mechanized, catches increased because their value exceeded the cost of mechanization. Because more fish were caught, the first signals of a general stock decline were noticed soon afterwards. Before that, fishermen had moved ever further from their home ports in order to maintain the high catch rates they expected. When the more distant catch rates themselves declined, fishermen became caught in a competitive rack from which the only escape was for some to stop fishing, to go out of business. This is the central theme of this book and, as will become clear, the process of industrialization continued until 1 January 1977, when the large distant-water vessels were excluded from the coastal sea of nations foreign to them. On that date many, but not all, nations declared their sovereignty over their Exclusive Economic Zones (EEZs) out to 200 miles from baselines.

In this chapter is given a brief history of the effects of the industrial revolution upon fisheries in England and of industrialization upon the first fisheries with steam-driven vessels in Europe and in North America.

A brief indication will be given of the consequence both at the time and later.

The effect of the industrial revolution on English fisheries

From the Suffolk shore in England, decked boats sailed to Iceland in summer for cod in the sixteenth century; after the discovery of cod on the Grand Banks in the early years of that century, fishermen from France, Portugal and Britain sailed there to catch them, as described in Chapter 3. During the eighteenth century, beam trawlers worked from Holland, Belgium and France. A beam trawl (see Figure 1(*c*), p. 2) was hauled by a roundabout capstan with a single thick warp (Figure 42). The roundabout capstan had been invented by Barking men; it took five men and two apprentices two to three hours to haul the trawl. With the trawl shot, a stiff breeze was needed to maintain a speed of 3.7 km h^{-1} over the ground (Figure 43). Towards the end of the eighteenth century, about forty decked sailing vessels or smacks (of 12–21 m long with tan sails) from Barking Creek in the Thames estuary ranged the Southern Bight of the North Sea and landed their catches at Billingsgate (Anon. 1921). There are various versions of the origin of the beam trawl from Holland. Cutting (1955) suggested that the beam trawl was common in bays and inlets of southeast England, having originated from the Zuider Zee in Holland. In 1616–17, John Farsby of Barking had been granted a certificate to trawl, but in 1631 Channel fishermen petitioned the Privy Council about the Barking men, who were using a 7 m beam trawl, which was subsequently banned. Alward (1932) wrote, 'when the Prince of Orange landed at Brixham he brought with him a large number of seamen who settled there and started trawling in Tor Bay'. Russell (1951) noted that

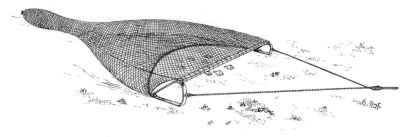

Figure 42. The beam trawl: the trawl is shown spread back from the beam supported on the iron trawl heads; it was towed on a single warp split into two bridles (Hardy 1959).

Figure 43. A smack at sea with the beam trawl lashed alongside and the cod end full of fish was hoisted on to the mast (Holdsworth 1874).

the word 'smack' is derived from the Dutch. Whatever the true origin of the beam trawl, the largest fleet in England towards the end of the eighteenth century was based in Brixham. In 1785, seventy-six smacks worked from Brixham in south-west England and they sent fish to London, Bristol, Bath and Exeter; the cost of transport to London was about £13 per load. Such was the state of the trawl fishery at the start of the industrial revolution.

Increased demand for fish in London in the early nineteenth century led to a migration of fishermen round the coasts. In 1812 Brixham trawlers moved to Dover because six small boats there had discovered an abundance of turbots on the Ridge and the Varne, banks in mid Channel. The price of fish fell in 1815 and in order to reduce costs the Brixham men migrated eastward and worked from Hastings, Ramsgate and Dover (Alward 1932). Between 1820 and 1840 there was a regular migration from Brixham to Dover from October to May (Dyson 1977). Between 1820 and 1829, the Brixham men discovered the New Bank (or Sandettié), the Falls and North Foreland grounds (all off the coast of Kent) where they found turbot, dory, brill and sole. In 1823, each vessel was landing up to 1000 or 2000 large turbot (Anon. 1921), a stock density which cannot be imagined today. By 1838, a trawler would have to tow for three tides for a single pair of soles (Alward 1932). Alward quotes John Cook as saying that hardly a sole or a turbot was to be found on these

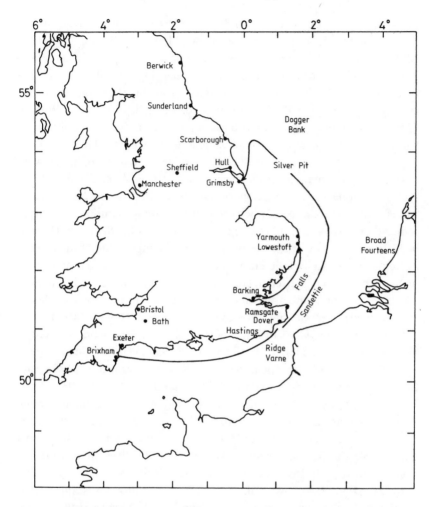

Figure 44. The two migrations: the first from Brixham to Ramsgate was extended to Scarborough, Hull and Grimsby; the second was from Barking creek to Gorleston.

grounds in 1840. This was the first migration (Figure 44(a)) and with it may have been associated the first records of depletion.

In 1830, the first fleet was organized by the fish merchants in Barking; they found a cutter and paid contract prices for daily delivery to Billingsgate. They worked the Broad Fourteens off the Dutch coast and probably subsisted on small plaice. They started to work northerly; in 1832 they moved to Yarmouth on the east coast of England and by 1835 to Scarborough on the Yorkshire coast (Cutting 1955). Soles were discovered in great quantity in the Silver Pit, south of the Dogger, during a cold winter, a 'Pits winter'; the date is given as 1834 (Alward 1932) and 1837 (Holt 1895). Cutting (1955) reported that local boats in that period from Flamborough had caught large quantities of soles. Dyson (1977) wrote that three or four Brixham trawlers working from Scarborough were dispersed by a storm in 1837. All save one reached port safely. William Sudds came in two days later with his net torn away by weight of fish; he had caught 2000 pairs of soles in a single haul. A particular cold winter appears in north-west Europe once every one or two decades in the twentieth century (or twice a decade during the Little Ice Age, Cushing 1982). An anticyclone remains stationary over Scandinavia and cold easterly winds blow over the North Sea, sea surface temperatures are sharply reduced and soles migrate to deeper water such as the Silver Pit (Woodhead 1964). The Silver Pit was named from the high value of the catches made during the first of the Pits winters. Holt (1895) noticed that in 1840, haddock curing had started in Hull; the fish were plentiful on the edges of the Dogger Bank and they may have been as important as soles in stimulating the second migration towards the north (Figure 44(*b*)).

Fleet owners Hewett's of Barking moved to Gorleston on the coast of Norfolk in 1851 and Hellyer's of Brixham in 1855 to Hull, which had been a whaling port. Alward, speaking to the Select Committee in 1893, said that a Devon accent persisted in Hull till that date (Anon. 1893). It was Holdsworth (1874) who distinguished the two migrations, the first from Brixham to Ramsgate and thence to Scarborough for the summer visitors and the second from Barking to Gorleston. Scarborough was a small port and in 1845 the smacks moved to Hull, but the facilities there were not very satisfactory. In 1848, the Great Northern Railway reached Grimsby, in 1852 the Royal Dock was built and in 1867 the pontoon was constructed on which the merchants could work. The Royal Dock in Hull was not completed until 1869 and St Andrew's Dock there was opened in 1883. The first migration from Brixham eastward was driven by economic factors, but the second may have been due to the depletion of grounds in the Southern Bight of the North Sea during the 1830s and 1840s as the fleets started to work from Barking creek. The existence of Barking as a fishing port ended when the fleets moved north in the second migration, perhaps the first consequence of excessive fishing pressure.

Fleeting had started in 1828. The fleet was commanded by an admiral. They shot their beam trawls at noon, towed for five or six hours and, as it took two or three hours to haul the trawl, the smacks made three hauls in a day. The fleet worked within 16 km of an anchored mark boat and some boats 'skirmished' a bit further afield. The admiral controlled the fleet by flag in the daytime and by rocket at night. When a smack left the fleet in the early 1850s, a flag was hoisted and the whole night catch of the fleet was put aboard. The fleets stayed at sea for six or eight weeks (Alward 1932) and a whole generation of fishermen spent their lives at sea with only six weeks in port every year.

The fish cutters (or carriers), were faster, more like yachts. They were introduced by fleet owners Hewetts and Morgans, and in 1864 the first steam carriers sailed from Barking, the *Hewett* and the *Lord Alfred Paget*. The steam carrier transported three times the cargo of a sailing cutter at three times the speed (Cutting 1955). Fish were transferred from smack to carrier by small boats in all weathers. 'Boarding', as it was called, to the carrier took place at the mark boat. Between June 1880 and December 1892, 4103 boys became apprentices, 1080 absconded and about 200 died or were drowned (Alward 1932). The reason for these appalling figures was that the small boats were manned by the boys, in the charge of a senior hand (Figure 45).

The Short Blue Fleet was founded by Alexander Hewett, and its well smacks for live cod (carried in sea-water tanks) made fourteen-week voyages to Iceland every summer; Robert Hewett said, 'My father once sighted Jan Mayen' (Alward 1932). Robert had the Short Blue Fleet clear

Figure 45. Boarding fish: the small boats carried fish from the steam trawlers and smacks to the steam carrier (Wood 1911).

the sea of boulders from which he built a mansion and a cottage hospital (Dyson 1977). In 1864, Hewett's Home Fleet worked the Southern North Sea and the Short Blue Fleet further north. Before the railway came to Yarmouth, the fish left the town at 5.00 p.m. to reach Billingsgate the following morning at 5.00 a.m. using forty-eight horses with twelve changes. The Home Fleet of 120–200 smacks would spend the spring off the Dutch coast of the Southern Bight of the North Sea, the summer off the north coast of Holland between Terschelling and Ameland, and in the autumn it would return to the Dogger and the Silver Pits. There were other fleets, Leleu's, Columbia, Great Northern, Coffee Smith's, Red Cross and the Gamecock; the last was shot up in 1904 on the Dogger Bank by a Russian battleship on the first stages of its voyage half the world away to disaster at the hands of the Japanese. In 1906 Hellyer's fleet comprised fifty-five steam trawlers and seven carriers. The last fleet sailed from Hull in 1928. As stock densities declined in the North Sea and as the steam trawlers searched further afield, fleeting was replaced by 'single boating'. As will be shown below, steam trawlers did not appear until 1881.

Before 1820, live fish caught by line had been brought to Billingsgate alive in well cutters, the holds of which were filled with sea water and live fish (Alward 1932). With the advent of fleeting, the smacks made long hauls, the fish were dropped dead on deck, and ice had to be used to pre-serve the fish. There were already ice houses and in centres of salmon fishery like Berwick-on-Tweed and salmon packed on ice were taken by fast carrier to London. Samuel Hewett first used ice from Norway, then from local farmers who flooded fields for the purpose and who sold it for 5–10 shillings per tonne; for some years it was their most profitable crop (Benham 1979). In the mid nineteenth century as many as 3000 people were employed near Barking. Alward (1932) quoted: 'the heavy frosts did not usually come till after Christmas, when we hoped to store enough ice until next November, but sometimes we bought Norway ice'. The ice house at Abbott Rd, Barking, had a capacity of 10000 tonnes 5.5 m underground and its walls were 2.5 m thick (Cutting 1955); the thin sheets of ice stuck together under pressure. Ice was first imported from Norway in 1859; much came from Lake Oppegaard where horse ploughs were used to harvest it (Alward 1932). However, Cutting (1955) records that the first ice came from Lake Wenham, USA, and the initial imports from Lake Oppegaard had to be labelled 'Wenham'. By 1880, 31132 tonnes year^{-1} were being imported, a quantity that increased to 61396 tonnes year^{-1} by 1893. The East Anglian Ice Company was founded in 1874 and, in the early years of the twentieth century, Norwegian imports declined. Cutting (1955) wrote that the fleet carriers used 5–16 tonnes of ice in winter and 10–20 tonnes in summer. During the age of fleeting, carriers brought in fish on ice every day and even with a longish voyage, the fish cannot have been more than two or three days on ice. When fleeting gave way to single boating, as steam trawlers became more mobile, the time spent by dead fish on ice increased to ten days in the North Sea and up to three weeks if the boats steamed to Iceland or the Barents Sea.

In 1849 the Manchester, Sheffield and Lincolnshire Railway came to Grimsby. Other railway systems developed and they owned the docks in many of the fishing ports, because they had built them. They developed the midland markets and ran express trains to London. They carried fish on ice from port to midland or London markets and, because of the rapid transit from smack to housewife in the days of fleeting in the 1850s to 1870s, fish was brought fairly fresh to the consumer. Only when fleeting was replaced by single boat trawling with longer voyages to more distant grounds did British people acquire a taste for fish that were more than dead and a smell of trimethylamine pervaded the fishing industry. Fortunately it dispersed when freezers went to sea in the 1950s and 1960s and children started to eat fish fingers.

Industrialization in the North Sea

The most important changes at sea occurred in 1860 when the *Heatherbell*, a paddle steamer tug from Sunderland, towed the smacks *Fearnot* and *Henry Fenwick* to sea, 8–16 km offshore. During the late 1860s, paddle tugs were using beam trawls from Sunderland and North Shields. In 1878, fifty-three paddle tugs were working on north-east coast grounds and, by 1884, twenty-four were at sea and in many of the marine pictures of the period there are paddle tugs portrayed on their way to sea (Figure 46). The first effective steam trawlers, *Zodiac* and *Aries*, were built in 1881 at a cost of £3000 and £3500, respectively (Dyson 1977); their steam capstans were worked automatically below decks. They were 34 m long, steamed at 16.6 km h^{-1}, burned 4 tonnes of coal per day and caught four times as much as a smack (Dyson 1977). They were independent of wind and of calm weather and, with their advent, fisheries had become industrialized.

An earlier steam trawler, the *Albatross*, had worked from Grimsby in 1856, but she could not make the working expenses (Alward 1911). But the successors of *Zodiac* and *Aries* spread rapidly from Grimsby, Fleetwood and Aberdeen. The development is described in the following table, showing the increment of steamers and a decrement of smacks.

Year	Numbers of steam trawlers	Average tonnage (tonnes)	Number of sailing vessels
1883	181	30.8	8443
1893	564	39.6	7369
1902	1573	56.3	5887

From Alward (1932).

In the 1880s, an iron steamer was used as a carrier, 30 m long with 30 tonnes of ice, 45 tonnes of coal and 3000 fish trunks; in 1881, she made 286 trips carrying 15 000 tonnes of fish (Cutting 1955). Within two decades, the effective fishing effort exerted by the British fleet in the North Sea increased enormously and the first signals of decline in stock density were noticed. Figure 47 shows the annual catches of plaice and haddock by four smacks from Grimsby between 1867 and 1892. The decline in plaice stock density is considerable, perhaps by a factor of 4; that of haddock is less marked and more variable. Indeed we might guess that there were good haddock year classes (fish born in a particular year) in 1880 and 1889. Between 1881 and 1914, the ports of Hull and Grimsby advanced whereas those of Lowestoft, Yarmouth and Brixham declined. Coal in the north

Figure 46. Paddle tugs were used to tow the smacks before the invention of the steam trawler (Dade 1933).

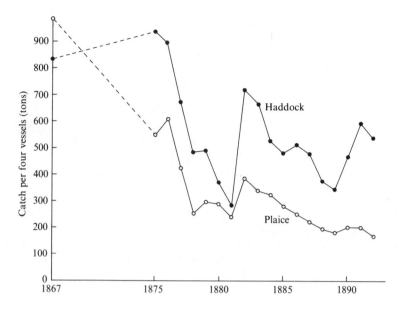

Figure 47. The decline in stock density, as shown by the annual catches of plaice
and of haddock by four Grimsby smacks between 1867 and 1892 (after Alward
1907). Broken lines indicate periods for which there are no data.

was cheaper, but the railways reached Hull and Grimsby long before they
could benefit Lowestoft and Yarmouth; in any case the lines from the
East Anglian ports were not laid conveniently to serve the Midland
centres of industry. The last smacks were working from Lowestoft as late
as 1938 as shown by the wonderful photographs of Ford Jenkins (1946).

In 1895, the otter trawl towed on two warps replaced the beam trawl
which had been hauled on a single warp. Scott of Granton had earlier
made the trawl to work with his patented doors, which spread the wings
of the otter trawl apart. James Alward of Grimsby invented the double-
barrelled steam winch which he put on deck. Then the two warps of the
trawl, one from each wing, were hauled independently over pulleys hung
from the gallows (or steel arches) both fore and aft. The rig is illustrated
in Figure 48. So, the trawl was hauled to the side of the vessel and was
lifted in until the cod end at the apex of the conical bag was dumped on
deck. This organization of gear on the deck of a steam trawler was to
remain the standard form until the invention of the stern trawler in the
1950s, in Britain. With this latter vessel, the trawl was hauled sıtraight up
the stern ramps, a much simpler arrangement. At first, only the larger
vessels were stern trawlers, but later it became clear that even for smaller

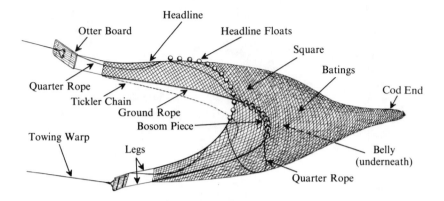

Figure 48. The otter trawl is spread by the otter boards and the headline is buoyed up by floats. The tickler chain stirs up the sea bed ahead of the ground rope. When the trawl is hauled, the quarter rope pulls the ground rope up to the head rope (Davis 1936).

boats such an arrangement was very efficient. By 1904, the otter trawl was used by steam trawlers from Hull, Grimsby, Fleetwood and Aberdeen.

Figure 49 shows the stages of exploration of the North Sea. Between 1825 and 1835, plaice, sole, turbot, brill, whiting and dab were being caught in the Southern Bight of the North Sea. By 1843, the smacks had reached the Silver Pit and during that decade they were working off the north-east coast of England. During the 1850s and 1860s English smacks were working off the German and Danish coasts and by the 1880s they had explored the Great Fisher Bank and the Scottish coasts. By 1881 the smacks had worked as far north as Shetland (Alward 1911) and by 1891, steam trawlers were taking plaice and haddock from Ingolf's Hoof (Ingolf's Hof de Hoek) off Iceland. The explorers were driven further afield from the ports because the local grounds were depleted. It was cheaper to steam further afield to find new klondykes, new gold fields, than to remain on nearer grounds where the fleets had worked. Thus each stage of exploration in the North Sea during the nineteenth century is really a measure of exploitation, of a progress towards overfishing.

The development of trawling across the North Sea was primarily one by the English smacksmen. Scottish fishermen depended upon a large herring fishery (as did the Dutch) and their liners caught cod. But when the steam trawlers were invented they were used at once from Aberdeen and of course the otter trawl itself was a Scottish invention. Beam trawls had been worked on the European coasts probably before they came to Brixham, but there was no development there until the steam trawler

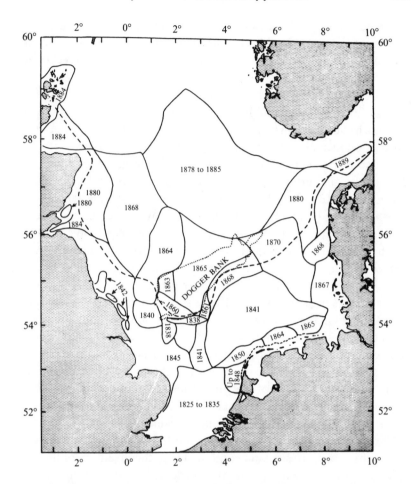

Figure 49. The exploration of the North Sea. The chart shows the stages of exploration and the text states where particular species were caught in the years indicated (Alward 1911).

appeared. Even then the rate of building was less than in England during the last two decades of the nineteenth century (Hérubel 1912). Another interesting point is that, although seven steam drifters (for the herring fishery) were built in 1882, the great increase in numbers was delayed until the first decade of the twentieth century (Kyle 1929).

I have already referred to the loss of boys at sea, probably when boarding. Dyson (1977) gives the numbers of men lost at sea between 1884 and 1902; it ranges from 152 to 492 men lost each year. But until 1880 a death at sea did not have to be reported unless the boat was damaged. However, at the International Fisheries Exhibition in 1883, the Duke of Edinburgh classified the annual losses at sea, 312 in England, eighty-six in Scotland and thirty in Ireland, and his report drew attention to the harsh risks of this dangerous trade. At Lowestoft during the 1950s, the port reckoned to lose one vessel each year.

The effect of the industrial revolution initially was to raise the demand for fish first in London and later in the Midland towns; at that time fish was much cheaper than meat, a condition which was maintained until the price of oil increased during the early 1970s. In 1851, there were a few fried-fish shops and in 1861, 300 persons were engaged in the trade; a 'gin drinking neighbourhood suits us best for people haven't their smell so correct there'. By 1907, there were 25 000 fish shops using 200 000 tonnes of fish (Cutting 1955), but today much of this by then admirable trade has faded. The first migration of fishermen after the Napoleonic wars from Brixham to Barking took place because the price of fish fell in the first years of peace. But the second migration during the 1820s and 1830s from Barking to Yarmouth and the Humber was perhaps a direct effect of Midland demand, expressed finally in the railway which reached Grimsby in 1849. During the mid century, fleeting became the natural method of fishing, with smacks at sea for eight weeks and cutters supplying the markets every day. Ice was culled as a crop at Barking and later imported from Norway to supply the steam cutters as they travelled from fleet to port. The initial effect of the industrial revolution was to stimulate demand, gather the fleets, import ice and instigate exploration northward in the North Sea as the nearer grounds became depleted.

The industrialization of fish capture started when the first paddle tugs towed the smacks to sea, and continued with the invention of the steam trawler with a steam capstan below decks. It was completed with Alward's invention (1895) of the double-barrelled steam winch on deck, which made the traditional steam trawler or 'sidewinder'. But the consequence of the industrialization of fish capture was that efficiency increased by a factor of 4, that stock density declined in the North Sea and that 'single boating' replaced fleeting (if rather slowly, with the consequence that vessels steamed further to find high catches and fish remained longer on ice before sale). With hindsight, all such effects were the consequence of reduced stock density. The fishermen were driven to work harder for less return in rougher seas. With the advantage of hindsight we see that fewer fishermen might have remained relatively prosperous in easier seas nearer their home ports. The reason for this disastrous drive to harsher seas was simple: a fisherman *will* spend his profit on another boat. Because the resources of the sea are finite and are now more limited than they have ever been, he should spend his profit on anything *but* another boat. However, had he followed this advice, the explorations would not have taken place then and all the distant waters might have remained unexplored, for a short time.

Evidence of stock decline

During the second half of the nineteenth century there were enquiries into the state of the fishing industry. There was the Royal Commission of 1863 (Anon. 1866), the Royal Commission on Trawling of 1885 and the Report from the Select Committee on Sea Fisheries of 1893 amongst others.

The 1863 Commission (Anon. 1866) was appointed to enquire into the sea fisheries of the United Kingdom and they answered three questions: '(1) whether the supply of fish from such Fisheries is increasing, stationary or diminishing; (2) whether spawn is damaged; (3) whether any existing legislative restrictions operate injuriously upon any of such fisheries.' There were no statistics recorded at that time, but records from the railways showed that 98 154 tonnes of fish were carried in 1862, 107 009 tonnes in 1863 and, in 1864, 120 454 tonnes. The first and third questions are those that concern us here; on the second, fishermen had believed that spawn was to be found on the sea bed. Most witnesses reported that catches had increased or had remained steady; catches had risen by a factor of 5 as demand for fish had risen (Samuel Decent). On the Dogger, they caught 6 tonnes in one catch, where earlier they had only caught 2 (Joseph Potter); hauls of 60, 70 or 100 baskets in one catch, very unusual fourteen years before (Richard Thomas Vivian). There was an abundance of fish at any time of year on the Dogger: 'They are finer fish, better fed' (William Isaac Markcrow). Samuel Hewett quoted the testimony of four skippers who caught 2–3 tonnes per haul on the Dogger. Christopher Spashett of Yarmouth also referred to average catches of 3 tons, with 140 score of haddocks. One witness referred to one catch on the Dogger of 326 'packages', or 57 tonnes, 'as much as we could get abroad' (Nicholas Apter). Between the English coast, Holland and the Dogger 'there was as much fish as ever there was' (George Pellett). Robert Cook of Ramsgate reported that 'they go fishing on the same ground that I was fishing on 48 years ago, and all the summer time you will see the trawlers trawling over a piece of ground not a quarter or a tenth of a mile broad; day after day, day after day, away go the trawlers over this piece of ground and they catch an abundance of fish'. The Commission, which included Professor T. H. Huxley, concluded that 'the total supply of fish obtained upon the coasts of the United Kingdom has not diminished of late years, but has increased; and it admits of a further augmentation to an extent the limits of which are not indicated by any evidence we have been able to obtain'. The conclusion was right, but the extension will find an echo subsequently.

However, some witnesses reported that catches of flatfish and soles had decreased (James Salter, Eliezer Johnson, William Holt). Robert Cook of Ramsgate said that catches of soles and plaice had fallen from 30 tons to 20 baskets. In Torbay, catches of hake, whiting, flatfish and soles had fallen in the ten years up to 1863 (Eliezer Johnson). Off the Start, hake and megrim were scarce and one pollack was caught each year, where earlier forty to seventy were taken each day. Thus, there was a little evidence of decline of flatfish vulnerable to the capture of immature fish in the eastern North Sea or of the decline of some species in the south-west where trawling had been practised for some decades. It would seem to a reader 120 years later that Huxley rode a little roughly over those fishermen who were already suggesting that bigger catches might result if small flatfish were no longer caught.

There was some evidence of decline in cod catches by the liners or 'codbangers'; Samuel Hewett told the Commission that he had '5 or 6 line vessels catching cod and whiting about the Well Bank. They make short voyages and bring the fish in five days. One vessel had over a hundred score . . . I have been giving them up, finding trawling pays better'.

There were two Acts in force (1 Geo I.c.18 and the Convention Act of 1843). In the first, all nets (except those for herring, sprat and pilchard) should have meshes of not less than 3½ inches (9 cm) from knot to knot and that nets of less than this size should be seized and burned. I suspect that this Act sought to control trammel nets, for trawls as we know them were not in widespread use in the reign of George I. But the Act forbade the landing or selling of fish less than a minimum size: brill and turbot, 16 in.; codling, 12 in.; whiting, 6 in.; mullet, 12 in.; sole, 8 in.; plaice and dab, 8 in.; flounder, 7 in. (1 in. = 2.54 cm). There were also limits to the length of beams on the beam trawls. The laws were not enforced at all and the Commission was very careful to establish that fishermen did not know of them and that nobody had responsibility for enforcement. The Commission recommended that all Acts of Parliament that profess to regulate, or restrict, the modes of fishing pursued in the open sea be repealed and that *unrestricted freedom of fishing* be permitted hereafter.

It is a truism that unenforced laws are an anachronism, but again the later reader looks a little wryly at the abandonment of regulation at a point in time when the first small signals of limited resources were being recorded. In the 1970s I was involved in efforts to limit the sizes of beam trawls. Minimum landing sizes would have demanded corresponding control of mesh sizes, the need for which was not properly established until the 1930s. But as will be shown below, by 1885 and 1893 the fishermen had become very concerned that the stocks of flatfish were being jeopardized by catches of very small fish in the shallow waters off Holland, Germany and Denmark.

In 1883 an international Fisheries Exhibition was held in London. Professor T. H. Huxley (then President of the Royal Society) gave the Inaugural Address and asked if the stocks of fish could be exhausted (Huxley 1884). First he pointed out that some fisheries could be exhausted: 'it is possible to net the main stream . . . as to catch every salmon . . . Pollutions may be poured into the upper waters of a salmon river . . . to destroy every fish in it.' At sea, however, 'the multitudes of these fishes is so inconceivably great that the number we catch is relatively insignificant; and, secondly, that the magnitude of the destructive agencies at work upon them is so prodigious, that the destruction effected by the fishermen cannot sensibly increase the death rate.'

Professor Huxley used the following argument. Shoals of cod in the Vestfjord, where they spawn in northern Norway, are 36–55 m thick and if the fish are a metre apart, there might be 55 million cod below one square kilometre. The Lofoten fishermen might take 30 million in an exceptionally good annual catch. If each cod eats one herring each day, that shoal takes 840 million herring each week; yet all Norwegian fisheries take less than 400 million. The average stock of the Arcto-Norwegian cod, which spawns in the Vestfjord, between 1961 and 1970 (Garrod 1977) was approximately 700 million. During this period, the stock of Atlanto-Scandian herring between 1950 and 1970 might have amounted to 4000 million (Murphy 1977). Huxley's numbers bear comparison with those made seventy to ninety years later: the shoal of spawning cod in the 1950s was about 32 km long (Bostrøm 1955) and in the 1880s the stock was lower than during the present century (Ottestad, 1969).

Huxley (1884) said: 'I believe, then, that the cod fishery, the herring fishery, the pilchard fishery, the mackerel fishery and probably all the great sea fisheries are inexhaustible; that is to say, that nothing we do seriously affects the numbers of the fish. And any attempt to regulate these fisheries seems consequently, from the nature of the case to be useless.' His first argument was exaggerated, although his sums were reasonable, but the extension was not, for two reasons (*a*) the cod and herring stocks were larger than any in the North Sea and (*b*) the exploitation in Norwegian waters was probably a small fraction of that in the North Sea. There was a little evidence of stock decline from the 1863 Commission, but the first real evidence in the North Sea itself had to wait the publication of the Select Committee's report in 1893 (Anon. 1893).

The Select Committee (of the House of Commons) of 1893 was appointed to consider the expediency of adopting measures for the preservation and improvement of the sea fisheries in the seas around the British Isles, including the prohibition of the capture, landing or sale of undersized sea fish, the prohibition of the sale or possession of certain sea fish during the period when their capture is forbidden, the fixing of close seasons, the prohibition or regulation of certain methods of fishing, the protection of defined areas and other like regulations, international or otherwise. Conferences of the trawl fishing industry in November 1888 and in April 1890 had drawn attention to the diminution of flatfish stocks probably due to the destruction of immature fish in the eastern North Sea; a further conference in 1892 asked for the introduction of minimum landing sizes and asked for the appointment of a Select Committee to consider the expediency of prohibiting the landing and sale of undersized fish.

G. L. Alward (Anon. 1893) said that the best catches had been taken in 1875 and that average catches had declined since then:

	tonnes voyage^{-1}	
	1875	1892
Prime (incl. soles)	4.06	1.52
Offal { Plaice	27.94	9.14
Offal { Haddock	50.80	40.64

Further the average size of plaice had decreased: in 1867 there had been thirty-six fish in each trunk, but this quantity had risen to fifty in 1892. John L. McNaughton in his evidence to the Select Committee reported that an average haul of 18 cwt (914 kg) of haddock had fallen to 10 (508 kg). A fleet admiral, William Crossley Normington, was asked about the capture of immature fish: 'you would get a certain amount of marketable fish from amongst the small, but the evil has been that the men would not select those; they want the lot.' If the catch of immature fish were prevented, 'of a natural consequence they are bound to grow larger, and if there was none to catch them they would have a chance to live'.

Alward in his evidence attributed the decreased catch per unit of effort to the greater catches of undersized fish; he was careful to distinguish them from immature fish, which the scientists did not do. More generally, it was the increase in numbers of steam trawlers from 1881 onwards which in addition used bigger trawls (with longer beams). If a steam trawler caught three times as much as a smack, the total fishing power was nearly doubled during that decade. Alward reported that in 1840 the average tonnage of vessels from Grimsby was 40 tonnes, but in 1892 the tonnage of the steam vessel had risen to 153. In 1835, the average beam was 11 m in length, 14 m in 1860 and 22 m in 1892.

A number of scientists gave evidence to the Select Committee, the most useful of which was presented by Ernest Holt. He was asked whether the outcry about the depletion of North Sea flatfish was justified and replied: 'It is perfectly justified, I believe; especially from what we are told on all hands of those eastern grounds, about which we hear as having formerly yielded an immense number of large fish: now, I can say for certain, there is hardly a large fish on them at all.' The importance of this evidence was that Holt was a scientist employed by the Marine Biological Association at Plymouth and he worked on the pontoon at Grimsby; in 1893 he was working at sea on the trawlers investigating the proportion of small fishes in the catches. Holt (1893–5) classified plaice landed at Grimsby by size and found that 28% by number were smaller than 33 cm, but this proportion had increased to 68% in May; he also reported that of thirteen hauls made in the eastern North Sea, 47½ baskets (out of 141) were discarded, the fish being about 10–18 cm in length. Alward had distinguished between undersized and immature fish, but Holt's study was really directed towards the immature, for obvious biological reasons.

The Select Committee recommended the adoption of minimum landing sizes for flatfishes, the extension of territorial limits (for fishery purposes only) internationally to protect immature fish, the establishment of an English Sea Fishery Board (like the Scottish one) and the improvement of the collection of statistics. No action was taken on these recommendations at the time.

The mackerel and menhaden fisheries on the eastern seaboard of the United States

The two pelagic fisheries off the eastern coast of North America were linked during the nineteenth century because both species were caught by purse seine in summer between Cape Hatteras and the Gulf of St Lawrence. Between March and June, the mackerel move north between the Chesapeake and the southern shoal of Nantucket about 30 to 80 km offshore and this northerly migration was followed by the fleet. During the summer and early fall the best catches were made between Cape Cod and the Bay of Fundy, often on George's Bank. The fleet was predominantly based in the States of Massachusetts and Maine, particularly from the ports of Provincetown and Gloucester.

The provident sea

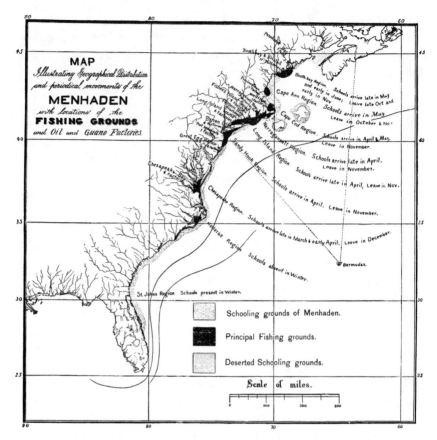

Figure 50. Chart of the menhaden fishing grounds (Brown Goode 1879). (1 mile
≈ 1.6 km.)

The menhaden fishery was described by Brown Goode and Clark
(1887):

As the menhaden appear in early spring in Chesapeake Bay and farther north
they increase rapidly in abundance until in some places the water is almost alive
with them. They prefer the shallow places along shore and in some years crowd
in great numbers into the brackish sounds and inlets and ascend some of the larger
rivers for a long distance until the waters become too fresh for them. They are
accustomed to swim in immense schools with their heads close to the surface,
packed side by side, and often tier above tier, almost as closely as sardines in a
box.

Figure 50 shows the fishing grounds of the menhaden with the more
important oil and guano factories. The two fisheries for mackerel and
menhaden took place at about the same season in the same region each
year.

Figure 51. The mackerel purse seine fishery; the schooner lies in the background and the fishermen haul in the seine (Brown Goode and Collins 1887*f*).

The origins of the purse seine are a little uncertain. Brown Goode (1879) writes that the first purse seine was made in 1826 by John Tallman I, Jonathan Brownell and Christopher Barker; it was 130 m long, 284 meshes deep and a 56 lb (25 kg) weight was used to hold it down. Unfortunately, they had some difficulty in retrieving the net. Brown Goode solved the problem of origin by saying that the first seines used north of Cape Cod were worked in 1850 by Captain Nathaniel Adams of Gloucester in the schooner *Splendid* and Captain Nathaniel Watson in the *Raphael*. These seines were 183 m long, 4 m deep with a 6.5 cm mesh, weighted down with 600–700 lb (270–318 kg). Eventually the mackerel purse seine reached a size of 400 m in length and 1000 meshes deep (Figure 51).

The mackerel seiners were 'swift sailers', broad beamed, and carried much sail. They had a clear deck, a bait box and a bait mill and the freshers carried an ice grinder; on the rail there was purse seine roller, 1.5–2.5 m long, 15–18 cm in diameter. The freshers would carry 175–200 barrels (22–25 tonnes) and the salters 175–500 barrels (22–63 tonnes). The seine boat towed and rowed well and was still in the water. At first it was a modified whale boat, 9.7–10.4 m long, of lap-streak style. In 1872, smooth-bottomed boats were built of battened seams sheathed with pine inside; they were faster, more durable and did not catch the twine of the seine.

There was a fresh market, with fish preserved on ice and a salt market for fish preserved in barrels. But there was also a canning outlet; in 1840, Charles Mitchell in Halifax, Nova Scotia, first canned mackerel for a limited market. In 1872, Edward Pharo of Philadelphia patented the packing of salt mackerel in small hermetically sealed tins; in 1879, Henry Mayo used tin cans of 5–10 lb (2.3–4.5 kg). Mackerel was also cooked in cans, with steam. The bulk of the catch during the 1870s was salted, but the real point is that all the outlets are of preserved fish, on ice, salt or in cans, and the material must have travelled for some distance into the industrial centres of North America as the continent developed after the Civil War. By 1879 the purse seine fishery was almost universal and in 1879/80 there were 468 vessels working.

The menhaden fleet comprised sailing vessels and steamers. In 1880, eighty-two steamers were working. They were screw steamers, 27.5 m long, with a 2.3 m beam with one mast forward and a crane for hauling in the catch. Each carried fourteen men – captain, mate, cook, engineer, fireman and nine fishermen (who were employed on the farms out of the fishing season). The purse seine was rather smaller than that used in the mackerel fishery; each steamer carried two. The vessels worked up to 16–24 km from the shore. The steamers displaced 60 tonnes, were made of hard pine with white oak frames and carried a central tank for stowing fish, which made the ship a safe vessel. The seine boat was really the same as that used in the mackerel fishery (Figures 52 and 53).

Originally the menhaden was used fresh, salt, as 'sardines' preserved in spices or for animal food, cod bait and mackerel bait. Large quantities were ploughed into the soil of the farms as fertilizer along the shores 'and filled soil with oil, parching it and making it unfit for tilling'. But the great outlet was for oil and guano. There are various sources for the origin of the factories, but it is likely that it was genuinely multiple. By 1877, there were fourteen factories in Maine, thirteen in Rhode Island, thirteen in Connecticut, thirty-one in New York, five in New Jersey and four in Chesapeake Bay. The fish was boiled in a large cooking vessel and pressed with steam presses; the oil was clarified in a shallow vat. The fish was carried from the steamer in 20-barrel (2.5 tonnes) cars on a sloping tramway into the factory, where it was dumped in large tanks, 50–75 barrels (6.25–9.40 tonnes) in each cooking tank. After steaming for half an hour, two thirds of the oil was released and the rest was expressed with 45–136 tonnes hydraulic presses. The large factories processed 3000 to 5000 barrels (375–625 tonnes) day^{-1}. Table 6 shows how important were the quantities of oil and guano.

During the decade 1873–82, the number of factories increased by one

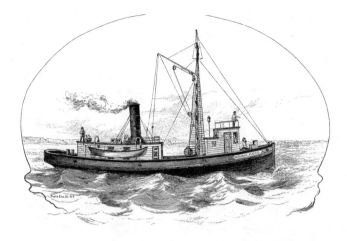

Figure 52. A menhaden steamer; the pilot house is forward and the dories hang from davits (Brown Goode 1879).

Figure 53. The menhaden purse seine: the steamer lies in the background and the net was hauled from the dories (Brown Goode 1879).

third, the number of sailing vessels was reduced by 45% and the number of steamers increased by a factor of 4. The number of fish taken reached a peak in 1880. Some 7.6–11.4 million litres of oil were made each year during the decade and 6352–72595 tonnes of guano. In 1880 more than half the catch was taken in New York State, Virginia, Rhode Is. and

Table 6. *Statistics of the fishery and of the manufacture of oil and guano*
(Brown Goode 1879)

Tables showing statistics of the manufacture of menhaden oil and guano in the
United States in the years 1873 to 1882. (Compiled from the Annual Reports of
the United States Menhaden Oil and Guano Association.)

	1873	1874	1875	1876	1877
Number of factories in operation	62	64	60	64	56
Number of sail vessels employed	383	283	304	320	270
Number of steamers employed	20	25	39	46	63

	1878	1879	1880	1881	1882
Number of factories in operation	56	60	79	97	97
Number of sail vessels employed	279	204	366	286	212
Number of steamers employed	64	81	82	73	83

Connecticut taking most of the rest. It was a considerable industry in
which up to 250 000 tonnes of fish were caught each year.

The oil was used primarily for currying leather in the tanneries because
of the scarcity of whale oil. It was also used for the lamps in miners'
helmets and in rope manufacture. A small quantity was used in lubri-
cation; it was also used to adulterate linseed oil in paint and in the manufacture of soap. It was sold in Boston, New York and New Bradford and
a fair proportion was exported to London, Liverpool and Le Havre.

In the United States, the industrial revolution started in about 1840. In
the next twenty years the population rose from 17 000 000 to 31 500 000,
and a firm domestic market developed for agricultural and domestic
goods. During this period before the Civil War, 260 cities were established and with them factories. After the war, the development of industry was resumed and the demand for oil increased. The greatest demand
for currying leather was perhaps partly associated with increased agriculture but also with increased communication from the railheads. Perhaps
it was also associated with the development of the American West.

There were various grades of guano: crude stock, half dry scrap, dry
scrap, dry ground scrap, acidulated fish, fish and potash salts and fish and
superphosphate. The finer grades compared favourably with Peruvian
guano and of course was useful for grass, grain and maize crops that need
nitrogen. Such was the consequence of the expansion of agriculture as
industry and population increased.

As will be shown in Chapter 13, fisheries developed in two directions:

the trawl fisheries for human consumption and the industrial fisheries for animal food. They first exploit white (or demersal) fish and second pelagic (or herring-like) fishes. In the North Sea, it was the trawl fleet that was industrialized in response to the rising demand of the increasing population. Off the eastern seaboard of the United States, an industrial fishery developed in response to more direct industrial and agricultural demand. But the salted mackerel supplemented the demand for protein. Thus, the two major trends of fisheries development throughout the twentieth century were revealed at the very start of the industrial revolution in the fisheries.

Although the first signs of depletion could be detected before then, overfishing followed the uncontrolled development of the steam trawler. The purse seine, which was invented in 1850, effective in the 1870s, was used throughout the world ocean on pelagic stocks; it became particularly efficient with the invention of the power block in 1957. It is extraordinary that the two gears most potent in the development of fisheries became effective within a decade of each other.

The Pacific halibut fishery (Thompson and Freeman 1930)

The halibut fishery off the north-west coast of North America was not large; the peak annual catch in the first decade of the century never exceeded 30000 tonnes. However, it arose when the railway reached the cities of Vancouver and Seattle and when cold stores were established to store the fish before transport to the cities of the east. Hence the fishery was a response to industrialization like that in the North Sea and that off the eastern seaboard. But the halibut stock suffered quickly from overfishing, stock density off Oregon and off Alaska having declined much as it had in the North Sea.

Early explorers such as Cook and Vancouver had noticed that the Indians landed relatively large quantities of halibut. The Indian hooks are particular in that a long barb between divergent arms was neatly fitted to the vertical mouth of the fish. Lines were twisted from cedar fibres, guts or sinews and from giant kelp and were as long as 160–200 m. the line was sunk with a weight and floated at the surface with bladders. The Indian fishermen worked usually in 20 or 40 m and a canoe with two men could watch up to fifteen lines. They could catch up to 100 fish per day from each canoe. The halibut were dried in smoke, as is still done by the Indians. It is possible that the total annual catch was of the order of 1000 tonnes.

The Canadian Pacific Railway reached Vancouver in 1885 and the

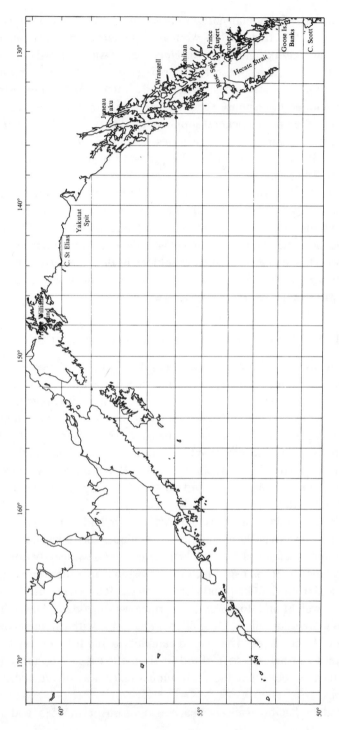

Figure 54. Chart of the ports at which were landed the Pacific halibut.

Northern Pacific Railway reached Tacoma on Puget Sound in 1887. There were various attempts to carry halibut on ice to San Francisco, but the main fishery started when catches were shipped east on the railways. In 1887 and 1888 vessels were fishing the Strait of Georgia and the Strait of San Juan de Fuca. In 1888 Captain Sol Jacobs of Gloucester, Mass., sent three schooners round the Horn to Puget Sound; they carried fourteen fishermen each and six dories. One of them, the *Oscar and Hattie*, landed fish from Cape Flattery, which formed the first shipment east of Tacoma. However, this venture failed, because the cost of transport was high. Other interests succeeded where Captain Jacobs had not, handling improved, transport rates fell, and by 1895 the Gloucester fishermen complained of competition with their eastern halibut fishery.

Figure 54 shows the ports where the fishery started. By 1900, schooners had started to work off Alaska in winter but returned to Cape Flattery in the summer. The early fishery from British Columbia was started by steamers which worked the grounds north of Vancouver Is. The first vessels carried fish caught by Indians to Vancouver; the first successful venture was made by Captain Freeman of the *Capilano* from Porcher Island, the fishermen working from small boats. The first cargo of about 5 tonnes, preserved in ice from nearby lakes, was shipped in boxes from New Westminster to New York. In 1895, dories were introduced and five or six companies were involved; in that year *Capilano* and *Coquitlam* landed about 400 tonnes, most of which was shipped east.

During the early period of the fishery, 'skates' of ten lines each, 100 m in length, were used and it was a winter fishery. The number of steamers increased from one in 1889, six in 1906, eleven in 1909 to thirteen in 1911, and many of the schooners were fitted with engines. The fleet gradually worked in deeper water and over a broader area. Schooners moved out towards Hecate Strait, Cape Scott, Banks Is., Rose Spit and Goose Is. Bank by 1903.

Halibut were caught off Alaska as early as 1895 but the fishery did not expand much until a cannery and a wharf were built at Petersburg. Then halibut could be shipped south to Seattle in boxes of 500 lb (227 kg) on ice taken from the glaciers. In 1901, there was an Alaskan fleet of twenty small schooners; at Icy Strait there were two steamers with four and three dories, respectively. Offshore exploration did not start until 1910 when the steamer fleet began to work off Alaska; in 1909 a cold storage plant was established at Ketchikan which allowed the fleet to fish throughout the year.

Up to 1909 or 1910, catches had built up as banks were discovered and exploited, but by this time fish in protected waters became scarcer. From

Rose Spit, Hecate Strait and Puget Sound, stocks were so reduced that it was no longer profitable to visit such grounds. Whenever a new bank was found, poor and discoloured fish were caught which were quickly eliminated. Because the halibut is very large and long lived, the average size declined quickly and at that time there was no market for small fish. The boats had then to search further afield, use more gear, more bait and more ice. From five halibut steamers working out of Vancouver between 1910 and 1915, it was shown that catches on formerly productive banks were falling at the rate of 75% per decade.

Technical advances were of some importance. Dories were introduced in 1895 which could be stacked on top of each other and so more could be carried on a steamer. In 1902 or 1902, slings were introduced on which fish were slung five at a time through the gills so that a dory could be unloaded more quickly. In 1902, Captain Joyce of the *Kingfisher* introduced landing nets, in which fish were thrown into the bottom of the dory, from which they were hauled directly on to the steamer. As consequence the fishermen were able to catch and handle twice as many fish per day. In 1903, auxiliary engines of 18.7–22.5 kW were introduced; from that time all new vessels were powered and larger vessels were used. By 1910, all but three of the Puget Sound fleet were powered.

The halibut was preserved on ice locally collected, and glacier ice was still in use in 1923 at Wrangell. The first ice plant was built in Vancouver in 1896; by 1905, there were a number of plants in Canada. In Seattle the first was built in 1892, and by 1903 four were at work. We have already noted one built at Ketchikan in 1909; in 1911 one was erected at Kildonan in Barkley Sound. The cold stores provided ice, frozen bait and space for surplus catches and their main effect was to steady prices. Thus the cold stores up and down the coast built to supply the eastern markets had other functions which served to make the industry more efficient. In Chapter 9 it will be shown that the cold store played another part.

Between 1902 and 1909, the average depth of capture was 80 m throughout the year. Between 1910 and 1915, the average depth of capture had increased to more than 160 m and in February that depth was as great as 260 m. Just as the boats steamed further away from Cape Flattery, so also they worked offshore in deeper and rougher waters. At about the same time, a boom in vessel-building reached a peak. Petrol engines were introduced in some vessels and some pelagic sealers joined the fleet after the end of that form of exploitation in 1911 in the sealing treaty.

In 1913, the spawning ground on Yakutat Spit was discovered either by the schooner *Idaho* or the steamer *Independent*. Later the banks between

Yakutat and Prince William Sound and Cape St Elias were discovered. Cold stores were built in Sitka, Wrangell, Juneau and Taku and a very large one with a capacity of 6250 tonnes and ice storage of 2000 tonnes was established at Prince Rupert. This was the western terminus of the Grand Trunk Railway and provided the main outlet for the Alaska fishery, which was three days steam from Seattle. The Canadians allowed American fishermen to land at Prince Rupert. A shift in the fishery then took place from the southern ground to Alaska and the frozen catch was carried eastward from Prince Rupert. In 1914 and 1915, the coast of Oregon was explored and about 1300 tonnes were taken in 1915, but catches were subsequently so low that the initial large hauls were probably taken from virgin stocks.

In 1915 various improvements were introduced on the vessels. Generating systems of 16 V were installed and, with deck lights, night fishing started. Power-driven anchor winches were put on the boats and the dories were hauled aboard by power. The holds became insulated. In 1913, lines were first shot over the stern and the long line was hauled over a roller with a power gurdy, or winch, in the waist of the vessel. So the dories were gradually abandoned by the larger boats. The number of fishermen on each boat was reduced and they were able to fish in rougher weather. Diesel engines were installed in 1921 and they spread amongst the new vessels built. Steamers were introduced in 1892, there were eighteen in 1913 and by 1929 there was only one left and all other vessels were diesel powered.

Between 1910 and 1930 stock density fell by a factor of 7, catches declined by a factor of 2.4 and effort rose by a factor of 3. Industrialization had led to depletion of local grounds, exploration of distant grounds, increased use of power and of developed gear and all these factors were driven by the persistent demand for frozen fish in the growing eastern markets.

The first industrialization of the fisheries

The industrial revolution took place in Europe and in North America but it occurred in fisheries in three somewhat limited regions, the North Sea, the region between Cape Hatteras and the Gulf of St Lawrence, and the Pacific north-west between Seattle and the Aleutian Is. The increase in population and the rise of the towns led to a demand for fish, cheap protein in Britain, oil for currying leather and frozen halibut for the new markets in Canada and the United States. The demands differed a little but the common consequence was an expansion of the fleet, smacks in the

North Sea, mackerel sailers off the eastern seaboard and small boats off British Columbia and Alaska. In the North Sea, large numbers of smacks were built, many purse seiners for the mackerel and menhaden fishery in the United States, and a small number of halibut liners in the United States and Canada. Fleeting was a phenomenon peculiar to the North Sea, but the drive away from local grounds to compensate for depletion was common to the halibut fishery and the North Sea trawl fishery; there was no evidence of such extended exploration in the mackerel and menhaden fisheries but the fish were very migratory and both species moved from Cape Hatteras to the Bay of Fundy. Pelagic stocks are vulnerable to recruitment overfishing (because recruitment declines under heavy fishing) and collapse suddenly, as did both mackerel and menhaden after the period examined here. Demersal stocks decline more in stock density under the pressure of fishing than do pelagic stocks, which would collapse under like pressure. Hence the search for distant grounds is a direct consequence of local depletion.

Fishermen quite reasonably strive to maintain stock density and they improve their gear, bigger purse seines, more dories, or more ticklers on the trawl. The most important innovation was the use of power, steam engines (and much later diesel), which made vessels independent of wind and tide and allowed them to reach market more readily. The double-barrelled steam winch on the steam trawler made that vessel with its otter trawl a much more effective fishing instrument. Such machinery is the traditional one in factories in the industrial revolution and the remarkable point remains that it came so late.

Ice became an essential ingredient of industrialization because, as stock density fell, vessels stayed at sea longer. In the menhaden and mackerel fisheries, ice was used in the small fresh fishery, but was used extensively by the fleet carriers in the North Sea. In the halibut fishery, ice was also used, but a considerable innovation was the use of cold stores in the ports. This was a function of the development of railways. The great symbol of the industrial revolution was of course the main agent of industrialization. In the North Sea the essential step was the arrival of the railways to Grimsby and Hull and in the Pacific north-west the trains came to Vancouver, Tacoma and Prince Rupert. On the eastern seaboard of the United States the railways were in place in the 1870s and 1880s and the factories scattered between the coast of Maine and the Chesapeake; they concentrated the catches and damped the input as did the cold stores in the Pacific north-west.

The most important consequence of mechanization and industrialization was the decline of stocks. Garstang (1903) showed that the catch per unit effort of the English North Sea trawl fishery declined by nearly half during the decade 1889–98, the period by which the steam trawler had become established. Although indications of depletion had been recorded by the 1863 Royal Commission and the Select Committee of 1893, Garstang's evidence was based on the whole English fleet, predominant at that time. In the Pacific halibut fishery between 1910 and 1931 stock density fell by a factor of 7 (Thompson 1952), a much severer decline, but in a smaller stock than that in the North Sea. These are the points of scientific evidence which will be discussed in Chapter 9. Our more immediate concern is the depletion felt by the fishermen as they tried to maintain their catch rates by steaming further afield. The overwhelming impression from the 1863 Commission was the contrast between low catches of many species off Brixham, of flatfish in the Southern Bight of the North Sea, and the high catches of cod and haddock on the Dogger Bank, recently discovered. Since then, fishermen throughout the world have steamed further, improved their fishing gear, in order to *maintain* their catch rates. As noted above, this was the drive that generated exploration within the North Sea, drew English steam trawlers to Iceland, the Barents Sea and Morocco in the first decade of the present century.

The demand generated by the industrial revolution affected fisheries fairly quickly, but industrialization came a little slowly, and persisted until the limit of resource became obvious. The stocks are limited, yet with care they can potentially release catches for long, perhaps limitless, periods. The transition from the free surging economy, as T. H. Huxley saw it, to a controlled one for the fish stocks is painful because some fishermen must leave the sea.

6

The great whales

Right whales and sperm whales were exploited by much the same methods from the twelfth century until 1873. Then Sven Føyn brought the faster whales, blue, fin and sei into exploitation with the steam-driven catcher and explosive harpoon. This was the industrialization of whaling, completed with the great floating factories in the 1920s in the Antarctic.

The history of whale fisheries is one of expansion followed by collapse. In this chapter is told the painful story of the International Whaling Commission. It was established in 1946 to conserve and manage the stocks of great whales. It failed to do so.

From the Biscayan fishery until 1946
(Scoresby 1820; Jenkins 1921; Spence 1980)

Introduction – the preindustrial fishery

As noted earlier (p. 7), Oppian describes a method of catching whales in antiquity. There are early records of whaling in Scandinavian, Norman and Anglo-Saxon literature. Before the development of the harpoon gun, the only stocks exploited were the right whales in the Bay of Biscay, right whales and bowheads off Spitsbergen, in the Davis Strait, Hudson Bay and in the Bering and Chukchi Seas, the southern right whale, the American gray whale and the sperm whales in subtropical seas mainly from latitude 40° toward the equator. The right whale (the right whale to catch) was passive, swam slowly, could be hunted easily in double-ended six-man rowing boats and floated when dead (some other species sank after death). There are two species, the black right whale (or *nordkaper* of the North Atlantic) and the Greenland bowhead; there were probably five stocks of the latter, Spitsbergen (in the region Spitsbergen, Greenland, Iceland and the Barents Sea), Davis Strait, Hudson Bay, Bering Sea and Sea of Okhotsk (Mitchell 1977). de Jong (1983) suggested that there were at least three stocks, Spitsbergen, west Greenland (including the Davis Strait) and Bering Sea (including the Sea of Okhotsk); he gave an account of life cycle and migrations based on old records and on the seasonal distribution of the ice. The sperm whale stocks may have had components north and south of the equator, with breeding seasons six months out of phase with each other; there is some discussion today on stock composition of sperm whales in the North Pacific.

The Biscay fishery

The Biscay fishery for the black right whale lasted from the twelfth century until the sixteenth and early seventeenth centuries. From mid June to December, whales were taken at the ports of Biarritz in France, and Caberton, Rasajus and Renteria on the north coast of Spain. Lookouts signalled the sighting of whales with drumbeat, bellringing or fires and then whaleboats were launched and the whales were run ashore. François Sopite invented the try-works on deck for boiling blubber at sea (Spence 1980; de Jong 1983). The whalers sailed offshore, probably in caravels, flensed the whale alongside and stored the blubber in casks (this was the essential process until the time of Føyn, see p. 152). Whales appear in the civic seals of Biarritz, Motrico, Guetaria, Fuenterrabia, Ondarroa, Lequeitio, San Sebastian, Saint Jean de Luz and Zubibara, all of which are towns in the Basque regions of France and Spain. Towards the end of the sixteenth century, fifty or sixty sail worked off Iceland from the Biscayan ports. They also worked in the Strait of Belle Island between Newfoundland and Labrador for the 'Grand Bay whale' (Mitchell 1977). The value of one cargo was worth two large galleons; here the try-works was established ashore and bones of the right whales have been found recently there (Tuck and Grenier 1981). They also reached Greenland and caught the bowhead, which they recognized as a different species (Mitchell 1977). Aguilar (1981) wrote that in the fourteenth and fifteenth centuries the Basque whalers ranged from the British Isles to Iceland, Greenland and Newfoundland.

Between 1517 and 1662, fifty-nine adults and thirteen young whales were taken in the port of Lequeito; between 1699 and 1789, twelve adults and six calves were landed in the port of Ciriquain. There were four sightings off the north coast of Spain in the nineteenth century and in 1977 one black right whale was seen off Vigo (Aguilar 1981).

A recent report to the International Whaling Commission (IWC) (June 1983) concluded that the Biscayan stock is near extinction, represented by no more than a few individuals.

The Dutch fishery off Spitsbergen (1611–1760)

Whalers always followed the explorers. Hull fishermen worked north from Iceland and traders to Archangel took walrus from the North Cape of Norway and Bear Is. (in the Barents Sea). In 1610, Captain Jonas Poole from the Muscovy Company saw whales off Spitsbergen and a possible base in Whale Bay. Whalers followed them, and in 1611 Captain Edge went to Elizabeth Bay on Bear Is. with six Biscayans. In 1613, six English ships sailed, guarded by the twenty-one-gun ship *Tigris* and, in 1614, fourteen ships from the Dutch Noordsche Company were accompanied by four thirteen-gun ships. By 1618 there were thirteen English and twenty-three Dutch vessels working off Spitsbergen. The Dutch venture was based on the European oil market; 'the abundance of Dutch whale oil caused a decline in the international price of seed oil'. But in Britain, rape oil was substituted for whale oil (Jackson 1978). At this time Grotius advocated the *Mare Liberum*, but James I supported the *Mare Clausum* (see Chapter 4). The two fleets divided the western coast of Spitsbergen between them. The English worked in the southern part from Bell Sound, Safe Harbour in the Isfjord, Horizon Bay and Magdalene Bay. The Dutch were based on Amsterdam Is., much further north. Later, the Hamburghers used Hamburg Bay and the French and Spaniards worked from the north coast of Spitsbergen. All fleets depended on the Biscayans for *specksioneers* (harpooners), coopers and *specksynders* (fat cutters). In 1623 the Dutch established a major shore station at Speerenburg (Blubbertown) on Amsterdam Is., where whales were plentiful close to shore. The animals were harpooned, towed to shore, flensed and blubbered, and casks of oil were then stored on the ships.

In 1626, two whalers from Zaandam caught a whale in the open sea and thus began a new venture; the separation of catching and processing increased the catch rate. In the 1630s, 200–300 ships, each with thirty crew, worked offshore and there were up to 18000 men in all at the settlement. Scoresby wrote that the 'bold and unconscious manner in which the whales resorted to the bays and sea coasts . . . their easy and expeditious destruction . . . , [and] immense herds in which the whales appeared in comparison to the number killed . . . , encouraged the hope that the profitable nature of the fishery would continue unabated'. These words might have been written at any time up to the middle 1960s. However, by 1640, the whales had become scarce in the bays, and between 1639 and 1645 the settlement at Speerenberg died out. de Jong (1972) distinguished three phases: 1610–70, a bay fishery; 1635–70, coastal and sea whaling;

Table 7. *Statistics of the Dutch northern whale fishery by decades,*
1669–78 (Scoresby 1820)

Decade	Vessels per decade	Whales	Barrels of oil		Whalebone (million lbs)		Vessels lost	Whales/ vessel
Spitsbergen								
1669–78	993	6414	425000	(70975)[a]	12.75	(5785)[a]	83	6.45
1679–88	1932	10919	590000	(98530)	17.70	(8031)	113	5.65
1689–98	955	4864	283600	(47361)	8.51	(3861)	82	5.09
1699–08	1652	8537	451800	(75451)	13.55	(2263)	62	5.17
1709–18	1351	4645	255700	(42701)	7.67	(3480)	51	3.44
1719–28	1504	3439	197400	(32966)	5.92	(2686)	40	2.29
1729–38	858	2198	130400	(21777)	3.91	(1774)	13	2.55
1739–48	1356	6193	289288	(48311)	8.68	(3938)	31	4.57
1749–58	1339	4770	202412	(33803)	6.07	(2754)	30	3.56
1759–68	1324	3018	146419	(24452)	4.39	(1992)	25	2.28
1769–78	903	3493	133977	(23374)	4.02	(1824)	31	3.57
Davis Strait								
1719–28	748	1251	111228				20	1.67
1729–38	975	1929	152791				14	1.98
1739–48	368	1162	79324				10	3.16
1749–58	340	513	46150				6	1.51
1759–68	296	818	53772				4	2.76
1769–78	434	1313	85396				8	3.03

[a] Numbers in parentheses are tonnes.

post-1670, the ice fishery. The last must have been a bowhead fishery, but it is possible that right whales were caught in the earlier phases (Mitchell 1977).

The period between 1645 and 1750–75 was one of Dutch supremacy. By 1670 nearly all Dutch ports were involved in whaling and by 1671 Dutch whalers were working in the Davis Strait, off East Greenland, off Iceland, Spitsbergen, Novaya Zemlaya and Jan Mayen. By the 1670s, there were persistent reports that whales were becoming scarce. Table 7 gives material for the period of Dutch supremacy (from Zorgdragers quoted by Scoresby (1820)) and shows the essential structure of the fishery.

During a period of 110 years, 58490 Greenland right whales were taken by the Dutch off Spitsbergen and 6986 in the Davis Strait. A combined total of 363466 tonnes of oil was brought back. Some 623 vessels were lost, involving nearly 20000 men; a proportion, however, would have been recovered by other vessels. Catches were also reduced by half or by one third. These statistics were the first published in any detail and were quoted in many publications.

A more detailed account is given by de Jong (1978, 1983), on the basis of a fuller search of port statistics and of log books. Figure 55(*a*) shows the number of vessels, Dutch and German, in the Spitsbergen fishery between 1661 and 1825. The fleet started to decline after 1770; between 1719 and 1796 the number of Dutch vessels is separated into those at east and west Greenland. Figure 55(*b*) gives the number of whales flensed between 1661 and 1823; from 1719 onward they are separated into catches west of Greenland or in the Davis Strait and east of Greenland. Except for a short period in the 1720s and 1730s the Davis Strait fishery comprised a small proportion of the Dutch fishery. Mitchell (1977) points out that the number of whales flensed is an underestimate of those probably killed. By the 1770s and 1780s, the number flensed was reduced by about a factor of 3. Figure 55(*c*) shows the stock density, or number of whales per vessel (from 1661 to 1822, separated by east and west Greenland after 1719), which had declined by a factor of 2 to 3. The estimates vary considerably from year to year, as if herds were found from time to time and perhaps destroyed or dispersed. If that were so, the whalers probably had to spend more time searching for their quarry; in that case the decline in stock density is underestimated (stock density is a subject of current discussion in the Whaling Commission). Figure 56(*d*) illustrates the decline in average size during the periods, as numbers of casks of blubber per whale flensed, from 1661 to 1824, separated according to east and west Greenland from 1719 onward. The decline is best seen as fairly continuous. de Jong (1983) makes the interesting point that both fisheries suffered between 1725 and 1735, probably a climatological effect. By 1750, the average size was smaller by one third and by the end of that

Figure 55. The Dutch Fishery. (*a*) Number of vessels, Dutch and German, in the northern fishery 1661–1825 (de Jong 1978); Dutch vessels, (1661–1729: ●), Dutch vessels at east Greenland (●) and west Greenland (▼) 1730–1825, and German vessels (■) (de Jong 1983). There are detailed statistics in Hamburg, but catches of whales and seals are not distinguished. Losses amounted to about 2–3% each year but in 1746, eleven were lost off east Greenland and twenty-nine in 1798. (*b*) Number of whales flensed between 1661 and 1823; from 1719, catches were separated into those from west Greenland (and Davis Strait) and east Greenland (de Jong 1983). (*c*) Number of whales per vessel (1661–1822): whole fleet (●); east Greenland (○); west Greenland (▼) (de Jong 1983). (*d*) Number of casks of blubber per whale flensed, an index of average size; east Greenland (●); west Greenland (●) (de Jong 1983).

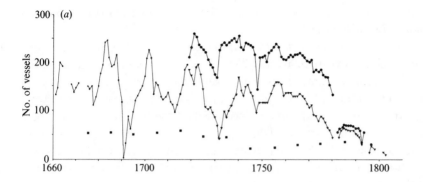

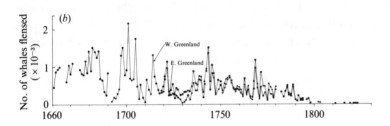

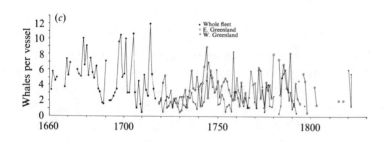

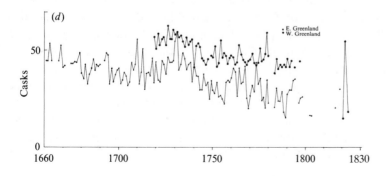

century the reduction was by nearly a half, but the east Greenland whales may have been larger.

The Dutch fishery started in the early seventeenth century and flourished between 1600 and 1760; stock density and average size both declined considerably. The early vessels, *fluits*, *galliots* and *hoekers* were cheap to build and Dutch interest rates were low; the companies were cooperative and their directors were called the book-keepers. In 1660, the vessels were 'doubled'; that is, the hull was doubled in thickness for the ice fishery. Such vessels were used only for whale catching and the blubber was brought back to try-works in Holland (de Jong 1978). The fishery became uneconomic in the second half of the eighteenth century in competition with those of other nations. Table 9 shows the apparent stock density in the British Spitsbergen fishery in the early nineteenth century, and it was much higher. Hence, the British ships were probably more efficient, perhaps in carrying more whalers. The Spitsbergen stock was probably reduced considerably at this time.

The American fishery before the Revolution

Boat whaling (1644–90) (Starbuck 1876)

The early settlers in New England probably observed Indian methods of whaling. They attacked right whales from small boats close to shore, dragged them ashore and cut them up there, although some initial cutting was done at sea. The try-works were, of course, built on shore. In 1639 an act was passed in Massachusetts to encourage fishing and whaling. In 1644, the first organized whale fishery was established at Southampton, Long Is. In 1669, twelve or thirteen whales were taken on the eastern end of Long Is.; Indian steersmen were often used. In 1688 the New Plimouth Colony made great profit from whale killing and Timotheus Vanderuen asked for a licence to fish the Bahamas and Cap Florida for sperm whales. In 1690 the New Plimouth Colony appointed inspectors of whales (who inspected before cutting began), to establish who owned the whale.

In 1690, the original purchasers of Nantucket employed Ichabod Paddock from Cape Cod to teach them how to kill whales and extract the oil. They erected a mast to search. From Folly House Hill, seeing whales at sea, they said: 'there is a green pasture where our children's grand-children will go for bread.'

Table 8. *The development of the
American whale fishery between
1762 and 1774*

Year	Vessels	Barrels of oil[a]	
1762	78	9493	(1585)
1763	60	9238	(1543)
1764	72	11983	(2001)
1765	101	11512	(1923)
1766	118	11963	(1998)
1767	108	16561	(2266)
1768	125	15439	(2578)
1769	119	19140	(3196)
1770[b]	125	14331	(2393)
1774[c]	360		

[a] Numbers in parentheses are tonnes.
[b] Clarke (1887*a*). [c] Spence (1980).

The whale fishery (1715–76)

In 1712, Christopher Hussey was blown offshore by northerly winds from Nantucket and he captured a sperm whale which he brought ashore alone. In 1715, six 30-tonne sloops put to sea for cruises of six weeks, with a few hogshead to carry blubber. By 1720, some boats were sailing north to Newfoundland and others south to the Gulf Stream; a small shipment of oil was sent to London. A harbour was built where the town of Nantucket now stands (Starbuck 1878). Each whaler carried two boats; sperm whales swim faster than right whales and the boats 'fastened on' after a strike in the 'Nantucket sleigh ride' (Spence 1980). They fished for right whales up to the end of May and turned to sperms between June and August (Dudley 1724). In 1730 there were 25 vessels of 38–50 tonnes each and they landed 3700 barrels of oil. By 1740, according to Starbuck (1878), the whalers sailed southward from Nantucket first and after 1 July they turned north for right whales off the Grand Banks. By 1746, brigs of 100 tonnes burden sailed for whales as far as the coast of Brazil, Guiana, the West Indies and Cape Verde. Table 8 shows the later development of the fishery.

During the Seven Years' War (1756–63) eight ships from Nantucket were captured by the French. But after that war, Britain's oil requirements for lighting (see below) were supplied by America (Spence 1980). Nantucket was the premier port, but not the only one; between 1752 and 1760, four vessels sailed from Martha's Vineyard. In 1760 there were three vessels from Long Is., four sloops from New Bedford in 1765 and,

in 1766, several from Rhode Is. A try-work was built at New Bedford in
1760. Such were the bases from which the much larger fishery was to
develop during the following century.

The whale fishery in the dispute between Britain and America
(1750–83) (Starbuck 1878; Clark 1887a)

In 1748, the British bounty became available to American whalers, provided they remained in the Davis Strait between May and August. During
the Seven Years' War, the British put an embargo on the Bank fisheries
because an expedition had been mounted to relieve Nova Scotia from the
French. In 1757, the whalers from Nantucket and Martha's Vineyard submitted a memorial asking for 'liberty to proceed with our voyages'. In
1761, a duty was levied on all whale oil and bone carried to Canada from
the new fishery for right whales in the Gulf of St Lawrence and the Strait
of Belle Isle. George Grenville abolished both duty and bounty in 1764,
in order to encourage the American fishery to replace the Dutch. The
increment in the American whaling fleet is shown in Table 8. Grenville
wrote: 'though we resign a valuable branch of trade in their favour . . .
yet the Preference is given upon truly national considerations, when the
inhabitants of America and of Europe are looked upon as one People'
(Jackson 1978).

In 1765, the Governors in the Gulf of St Lawrence and the Strait of
Belle Isle tried to exclude American vessels, with a series of petty restrictions, and the whalers left the region. Despite pressure from London,
Governor Palliser in 1767 reaffirmed the restrictions. On 10 February
1775 a Bill was passed in parliament which restricted the trade and commerce between Massachusetts, New Hampshire, Connecticut, Rhode
Island and Great Britain and the West Indies; the colonists were prohibited from any fishery on the North American coast, including the
Grand Banks. However, the island of Nantucket was relieved of the worst
features of the Bill.

Of this Bill, Edmund Burke in the Commons said (of issues broader
than those of the whale fishery):

when I know that the colonists in general owe little or nothing to any care of ours,
and that they are not squeezed into their happy form by constraints of a watchful
and suspicious Government, but that, through a wise and salutary neglect, a
generous nature has been suffered to take her own way to perfection; when I
reflect upon these effects, when I see how profitable they have been to us, I feel
all the pride of power sink and all presumption in the wisdom of human con-
trivances melt, and die away within me. My rigour relents. I pardon something to
the spirit of liberty. (Quoted by Starbuck (1878))

Subsequently Bills were passed in all colonies cutting supplies from the plantations to the British fleet.

During the war, all prisoners from American vessels, often whalers, had to serve as common sailors on His Majesty's Ships; John Adams, when in Paris (where he was negotiating the preliminary peace treaty between America and Great Britain), noted that, in the English fishery off the Plate, all officers and sailors were American (see Starbuck 1878). In 1778 the French tried to negotiate a treaty giving them the right to fish the Grand Banks. Samuel Adams (John Adams' cousin and a member of Congress) said: 'I hope that we shall secure to the United States, Canada, Nova Scotia, Florida too *and the fishery* by our arms or by treaty' (quoted by Starbuck (1878); my italics). In 1781, a British admiral, Digby, allowed twenty-four vessels to work from Nantucket, but several were taken by American privateers. In the same year the Nantucketers submitted a memorial to the General Court of Massachusetts asking for 'free liberty both from Great Britain and America to fish without interruption'. The treaty between the United States and Great Britain included 'the freedom in the fishery off Newfoundland'. But the American whaling fleet had disappeared. Throughout its history the whale fishery was profoundly affected by wars and their consequences.

British fisheries

There were two major British fisheries, the northern one for the bowhead whale off Spitsbergen and in the Davis Strait from about 1733 to 1824 and that for right whales and sperm whales south of the equator between 1775 and 1849.

The northern fishery or the Greenland trade (1733–1824)

In the first half of the eighteenth century the rapidly growing woollen industry in England consumed oil to clean the wool; 3 gallons (13.64 l) of rape were needed to clean 100 yards (91.44 m) of white cloth. By the 1740s, the streets of London were lit with whale oil and this practice spread to other towns. The new factories were lit by oil lamps. Upper-class houses were lit by spermaceti candles imported from the new American sperm whale fishery (see below). Whalebone was used in skirts, bodices and stays. Between 1725 and 1750, the Dutch industry started to decline, partly because of falling catch rates and partly because the continental rape oil seed industry was expanding. Further, American whale oil was cheaper (Jackson 1978).

In 1748 a bounty was introduced to encourage British whaling and in 1759 it was increased. The Dutch fishery had probably suffered during the Seven Years' War (1756–63) and the harpooners, line managers and steersmen in the British fishery were mainly Dutch. From 1762 to 1765, the fishery declined somewhat in competition with American imports (Jackson 1978).

During the American Revolution (the War of Independence) most of the American whaling fleet, more than 300 vessels, were captured or destroyed. Whalers from Nantucket and Rhode Is. were dispersed to Newfoundland, Canada and the Falkland Is. (Jackson 1978). In 1783 a duty of £18 ton^{-1} on whale products imported into Britain deprived the Americans of their trade (Jenkins 1921); before the war, the oil had been worth £30 ton^{-1}, but in 1783 the price fell to £17 ton^{-1}. Yet a price of £25 ton^{-1} was needed to make a profit. John Adams, the American minister in London after the war, wrote: 'we are surprised that you prefer darkness and consequent robberies, burglaries and murders in your streets to receiving, as a remittance, our spermaceti oil' (see Starbuck 1878).

In 1785, William Scoresby sailed from Whitby in the *Henrietta*. By 1788, there were 250 vessels, many of which were of American origin, but which had been pressed into service in the Royal Navy. They were three-masted square-riggers, of 400 tonnes, many of them double or treble planked against the ice, with four to seven whalers carried amidships. Hence they could sail earlier and bore through the dense drift ice south of Spitsbergen to reach the whales (de Jong 1983). From the British ports, the whalers sailed up the east coast of Britain and completed their crews in the Orkney and Shetland Is. In March and April, they sailed to the ice north of Spitsbergen and in it they drifted westward towards the coast of Greenland for four or five months. To reach the Davis Strait they sailed in February and worked in three areas: 65°–66° N, 68°–69° N and 71° N. By the end of the eighteenth century, whales were scarce enough for ships to go for weeks at a time without seeing a whale (Jackson 1978).

In 1807, Scoresby invented the protected crow's nest and, by 1812, 6000 men were working from 138 ships (Jenkins 1921; Spence 1980). By 1816, rape oil was cheaper than whale oil; prices and duties alternated for a few years; in 1822 the duty on rape oil was reduced; in 1824 the bounty was ended. Between 1814 and 1823, coal gas lighting was gradually introduced; up to 47 gasometers were built in London. By 1842 the number of whalers was reduced to 15% of the number between 1815 and 1819. The causes of decline were a combination of economic factors with reduced catch rate (Jackson 1978).

Table 9 shows the catch statistics of the British northern fishery

Table 9. *Statistics of the British northern fishery 1815–42 (Jackson 1978)*

Year	Spitsbergen vessels	Davis Strait vessels	Total	Number of whales	Tuns of oil[a]	Ships lost	Whales per vessel
1815	98	48	146	733	10682	1	5.02
1816	101	45	146	1330	13590	1	9.11
1817	97	53	150	828	10871	5	5.52
1818	94	63	157	1208	14482	2	7.69
1819	96	63	159	988	11401	12	6.21
1820	102	57	159	1595	18745	2	10.03
1821	80	79	159	1405	16853	14	8.84
1822	61	60	121	630	8663	8	5.21
1823	55	62	117	2018	17074	3	17.24
1824	32	79	111	761	9871	1	6.86
1825	21	89	110	500	6370	5	4.55
1826	5	90	95	512	7200	5	5.39
1827	16	72	88	1162	13186	1	13.21
1828	14	79	93	1197	13966	3	12.87
1829	1	88	89	871	10672	4	9.79
1830	0	91	91	161	2199	19	1.77
1831	8	80	88	451	5104	3	5.13
1832	19	62	81	1563	12610	5	19.28
1833	3	74	77	1695	14508	1	22.01
1834	7	69	76	872	8214	3	11.47
1835	1	70	71	167	2623	6	2.35
1836	3	58	61	70	707	2	1.15
1837	15	37	52	122	1356	2	2.35
1838	31	8	39	466	4346	1	11.95
1839	29	12	41	115	1441	0	2.80
1840	11	20	31	22	412	2	0.71
1841	11	8	19	52	647	0	2.74
1842	14	4	18	54	668	0	3.00
				21548	238460		

[a] 1 tun = 100 kg.

between 1815 and 1842. Until 1820, about 100 vessels worked off Spitsbergen and subsequently numbers declined. The fleet in the Davis Strait increased between 1815 and 1825, vessels remained about eighty in number until after 1832, after which the fleet diminished in numbers. The total catches of whales are combined for the two stocks, so estimates of stock density are confounded. But in spite of that, the number of whales caught per vessel declined sharply by a factor of 3 between 1835 and 1842. The implication is that the stock off Spitsbergen was further reduced and that the Davis Strait stock was also reduced. A total of 21548 whales was caught (much fewer than had been taken earlier by the Dutch) and they were boiled down to 238460 tuns of oil (about 24000 tonnes). Vessels were lost every year and in some years (1819, 1821 and 1830) a significant proportion was lost.

The English trade in northern waters declined during the 1840s. But Scottish ports continued to send whalers to northern waters. Between 1848 and 1857, they landed on average 111 whales per year and 95917 seals per year; it was the seal capture that sustained the vessels. Between 1861 and 1878 the premier port, Peterhead, gave way to Dundee, but the average annual number of whales fell from 38.2 in 1861–4 to 20.4 in 1875–9. Later still, the number of whalers out of Dundee declined:

Year	Whalers year^{-1}	Tuns boat^{-1}	Whales year^{-1}	Whales boat $^{-1a}$
1875–9	12.6	54 (5.4)^b	61.5	4.58
1880–4	10.4	76 (7.6)	71.4	6.87
1885–9	10.0	36 (3.6)	32.5	3.25
1890–4	7.4	46 (4.6)	30.8	4.16
1895–9	6.4	40 (4.0)	23.1	3.61
1900–4	5.8	32 (3.2)	16.8	2.89

^{a}Assuming 11.07 tuns whale^{-1}.
^{b}Values in tonnes given in parentheses.

Thus stock density, as tuns per boat, declined by about half and catches declined with fading effort (Jackson 1978).

The Dutch, the British and the Scottish fisheries for the Greenland right whale each ended for economic reasons. But their combined effect reduced the stocks off Spitsbergen and in the Davis Strait (where, as will be shown below, American fleets had an effect) to very low quantities. The fisheries ceased at the turn of the century and there have been sporadic reports of sightings. Jonsgård (1964; 1981) reported seven sightings (observed by Norwegian sealers) of Spitsbergen bowheads since 1962 between north-east Greenland and Novaya Zemlaya, five of them in the late 1970s. Mitchell (1977) believes that the initial stock of the Spitsbergen group amounted to about 25000 animals and the Davis Strait group to 6000.

The southern fishery (1790–1843) (Jackson 1978)

In 1775, Samuel Enderby from Boston, Mass., arrived in London with Alexander Champion and John St Barbe; in the following year fleet

Table 10. *Average annual*
number of vessels clearing
British ports for the
southern fishery
(Jackson 1978)

Year	Vessels
1776–80	11.0
1781–5	10.8
1786–90	42.2
1791–5	42.8
1796–1800	28.8
1801–5	39.4
1806–8	25.7

owners Enderby's grossed £6676 from their whaling activities. In 1786, the Shipping Registration Act was passed in parliament and, as a consequence, captured American whale vessels were distributed amongst the British. The same year saw the end of the Nantucket dispersal (see below) to the benefit of the London trade. At this time vessels worked in the South Atlantic between 7° N and 36° S, possibly for southern right whales; they were smaller than those in the Spitsbergen trade, and oil was boiled down in sheltered bays, so the casks contained oil, not blubber. But the magnet was the reports of sperm whales in the Pacific made by Captain Cook (Barrow 1906).

The East India Company had a British monopoly in the Indian and Pacific oceans. In 1788, one whaler rounded Cape Horn and then the Cape of Good Hope. In 1789, Enderby reached Botany Bay. In 1791/2, ten vessels sailed east of the Cape of Good Hope and eighteen west of Cape Horn searching for sperm whales. In 1795, 'southern premiums' were increased because of the demand for good-quality lamp oil. In 1803, bay fisheries were established for the Tasmanian right whale, and as late as 1841 thirty-five were caught. Between 1808 and 1815, the southern fishery was disrupted by the Napoleonic wars. Table 10 gives the average number of vessels clearing for the southern fishery between 1776 and 1808. The fleet was reduced by half between 1820 and 1825; there were only nine ships in 1841–3. By 1835, there were more Australian vessels than English and finally, in 1849, Samuel Enderby sailed for New Zealand. By 1818 200 American vessels were working in the Pacific on sperm whales. They no longer depended upon an export trade to Britain

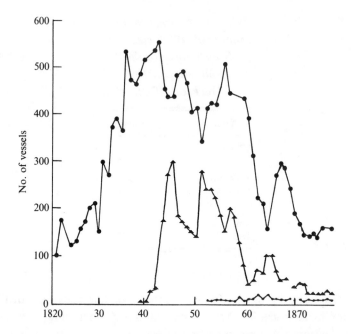

Figure 56. Number of vessels in three American fisheries; sperm whale in the Pacific (●), bowhead in the Bering Sea (▲), and bowhead in the Davis Strait (●) (Clark 1887*a*).

for they supplied oil for the pioneers' lamps as they trekked towards the western United States. The British had had a monopoly in the north and suffered competition in the south.

The American sperm whale fishery 1784–1880

In 1785 the United States introduced a bounty of £5 per tonne for white spermaceti, £3 per tonne for yellow spermaceti and £2 per tonne for whale oil. During the Revolution, people in America had become used to tallow candles. But in 1789 a revival started and in 1791 the price of sperm oil increased; a few vessels sailed from Nantucket and one from New Bedford. In that year the first whaler reached the coast of Chile. In 1800 the whalers reached Peru and the Line, as far as the Galapagos Is. They saw 'vast herds of cachalot' (the sperm whale). In 1802–3, they reached the China Seas and the Moluccas and in 1818 the Offshore Ground was discovered in latitude 5°–10° S and longitude 105°–125° W. French privateers had caused trouble in 1798–9, and between 1812 and 1815 war with the British stopped American whaling except from Nantucket.

There were two routes to the Pacific, the first by Cape Horn and the second by St Helena, Tristan da Cunha to the Cape of Good Hope, the Seychelles, New Holland, Australia and New Zealand (Spence 1980). In 1820–2 the Japan ground was discovered, which remained the principal area of activity until 1839. The high point of prosperity was reached in 1857 when 5 329 130 gallons (20 173 985 l) of sperm oil were landed. At this time the whalers worked off Madagascar, the Gulf of Aden and the Seychelles and they passed through the Straits of Timor to the Japan ground, whence they returned rounding the Horn.

Clark (1887a) gives the total number of vessels in the whaling fleet from 1821 to 1880, the number in the North Pacific from 1835 to 1880 and the number in the Davis Strait and Hudson Bay fleet. Hence, the number of vessels can be allocated to each of the three fisheries (Figure 56). Figure 57 shows the mean number of sperm whales caught in each decade between the first decade of the nineteenth century and the first decade of the twentieth, together with the number of American vessels lost (Best 1983); this estimate was derived from the quantity of oil landed adjusted by the average catch per year and average length of voyage (Townsend 1935). The total catches of sperm whales started to decline in the 1850s and 60s and subsequently continued to do so. The average catches in each decade were:

Year	Sperm whales vessel^{-1}	Vessels
1821–9	28.7	155
1830–0	13.7	374
1840–9	9.6	481
1850–9	7.0	419
1860–9	7.3	287
1870–9	9.8	173

The British, French, German, Dutch, Australian, Chilean and New Zealander whalers also took part in the fishery (Best 1983). The stock density (catch per vessel) declined during the century by a factor of 3. During the first thirty years, when the major decline occurred, the number of vessels (the fishing effort) increased by a factor of 3. Catch and stock density declined at the same time so the stock of sperm whales suffered from recruitment overfishing. Subsequently the fleet diminished, and in the last decade stock density may have recovered a little. From Figure 57, a peak catch of over 50 000 per decade was reached in the 1830s, after which it declined. Tillman and Breiwick (1983) gave preliminary estimates of stock in the nineteenth century.

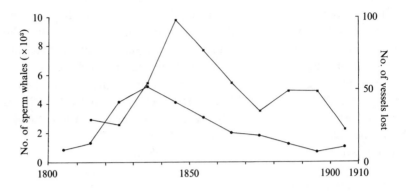

Figure 57. Number of sperm whales caught by decades between 1800 and 1910 (■); number of vessels lost (●) (Best 1983).

During the American Civil War (1861–5) the Confederate privateers *Alabama* and *Shenandoah* destroyed many vessels from New England. However, Starbuck (1878) recorded many desertions in the latter days of whaling. Many ships were fired by mutinous crews and many crew signed on merely to sail to the Californian goldfields. Starbuck attributed the collapse of the fishery to 'scarcity and shyness of the whale', extravagance in fitting out, the character of the men employed and finally to the introduction of coal oils.

The North Pacific bowhead fishery (1847–1936)

There were probably two stocks exploited in the North Pacific, right whales in the Sea of Okhotsk and bowheads in the Bering Sea, Beaufort Sea and Chuckchi Sea. The exploitation of the first started about 1847 (Mitchell 1977). In 1848, Captain Thomas Roys sailed through the Bering Strait and found the bowheads, 'oil rich, baleen laden and docile'. Figure 58 shows the annual catches of bowheads from 1848 to 1936 (Marquette and Bockstoce 1980; Bockstoce 1980); statistics were derived from a painstaking analysis of the whalemen's logbooks. However, the whale records do not distinguish between Okhotsk and Bering Sea stocks (Mitchell 1977). The figure shows a rapid rise to a peak catch after four years, with a subsequent dramatic and maintained decline. The fleet probably stopped fishing during the 1880s; during the whole period, the Eskimos caught bowheads and it is their catches which are recorded in the twentieth century. During the 1850s up to 200 vessels sailed through the Bering Strait, but subsequently the numbers were sharply diminished.

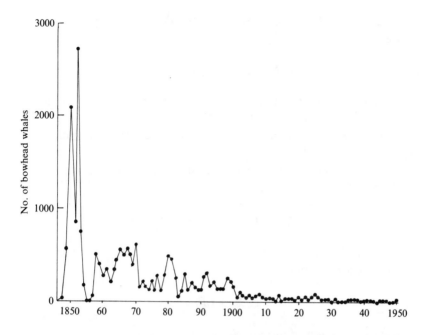

Figure 58. Annual catches of bowhead whales from 1848 to 1950 in the western Arctic Ocean (Marquette and Bockstoce 1980; Bockstoce 1980).

During the high period of the North Pacific fishery, the whalers would leave New England in autumn, round Cape Horn in the southern summer and reach San Francisco or the Hawaiian Sea in April and leave for Hawaii or San Francisco in October. One of Townsend's (1935) charts shows the migration of whales from the Sea of Okhotsk in early summer through the Bering Strait. In 1866, the whalers turned to walruses and gray whales. In 1880, the price of whale oil fell and that of baleen rose; steam vessels were used and they worked off the Mackenzie (Bockstoce 1980).

Because the stock of Bering Sea bowheads has been studied in some detail, it is relatively well known and models have been constructed to describe its history. Population models of whales start with the stock before exploitation, because catches have always been recorded quite well from the start.

With an earlier estimate of catches, Breiwick, Mitchell and Chapman (1981) have estimated the initial stock of Bering Sea bowheads. In 1978, 2264 bowheads were counted off Point Barrow and this provides an estimate from which calculation can proceed backwards. The following equations were used to describe the dependence of recruitment on parent stock:

$$r_t = M + (1 - P_{t-\tau}/P_0)(r - M)_{max}$$ (1)

where r_t is the recruitment rate; $(r - M)_{max}$ is the maximal net recruitment rate; P_0 is the initial stock in 1848; M is the natural mortality rate; τ is the lag time for population response; $P_{t-\tau}$ is the stock at the beginning of the season $(t - \tau)$; t is the time in years.

The relation (Allen 1966) between stock in the year $t+1$, P_{t+1} and that in the year before, P_t, was given by:

$$P_{t+1} = (P_t - C_t)\exp(-M) + R_t$$ (2)

where C_t is the catch in numbers in the year t; and the gross recruitment between seasons t and $t+1$,

$$R_t = r_t P_{t-\tau}.$$

Equation (2) is solved recursively.

Tillman, Breiwick and Chapman (1983) repeated the calculations with new estimates of total kill. In the model $(r-M)_{max} = 0.01, 0.03$ and 0.05; $\tau = 5$ and 7; the 1977 stock level = 1500, 2100 and 2700. The range of estimates for the stock in 1848 was 9300–16400. Breiwick and Mitchell (1983) used the same model with an additional lag of three years and 1978 stock levels of 1800, 2400 and 3000. The stock estimates in 1848 ranged from 9000 to 18000. The kill has risen in recent years to satisfy the demands of Eskimos in Alaska. In an earlier paper, Breiwick *et al.* (1981) suggested that, with different catch estimates, an annual kill rate of 30 year^{-1}, the recovery of the stock would take a long time, 75–153 years depending upon the magnitude of the parameters. With a smaller catch, of course, recovery would be quicker. If the catches includes bowheads from the stock in the Sea of Okhotsk, then the stock in the Bering Sea was overestimated.

In the study of whale populations, this paper is of some importance. Catch records are good and, although the estimate of Eskimo kills are uncertain, they are low. The five stocks of bowhead have not yet recovered from their earlier exploitation, the Biscay whale is no longer seen off St Jean de Luz and there are no longer herds of right whales in the bays on the western coasts of Spitsbergen. It is possible that at low stock levels, the reproductive rate is reduced, if the chance of encounter between male and female is less.

The Davis Strait fishery (1853–80)

Whalers from New England worked in the Davis Strait before the American Revolution, as described above; indeed, in 1744, 183 whaling vessels sailed to the Strait. Between 1853 and 1880, the number of vessels in any one year was never greater than seventeen. The fishery took place during a period when the sperm and bowhead fisheries still involved more than 200 vessels. Mitchell and Reeves (1981) believe that the present stock of perhaps 600 survive from an initial stock of 11 000 in 1825.

Conclusion

Between the period of the Biscay fishery and the end of the American fisheries, the method of capture did not change much, but vessels became larger and the number of whale boats carried increased; hence an increase in fishing effort may well have been unrecorded. Across the countries the number of whales caught per vessel may give an underestimate of the decline of stock density.

The need for whales changed across the centuries. It was first the general requirement for oil in competition with rape oil. As the Industrial Revolution started in England in the eighteenth century, the English used whale oil to clean their cloth and light the streets of London. In the following century, sperm oil was used to light no longer rich men's houses but the lamps of American pioneers, and whalebone was used in women's clothes.

The stock of sperm whales was reduced but it survived to be exploited again today. The stocks of right whale and Greenland whale declined so far in the northern hemisphere that their recovery rate from today must be slow, if indeed it is positive. The bowhead survives in the North Pacific and the Pacific gray whale can be seen as you flip from Los Angeles to San Diego; indeed, because the numbers have well recovered, the International Whaling Commission has allowed aboriginal capture of 178 gray whales and some bowheads. Of the other stocks we shall not know until the Greenland whales are seen in numbers in Hudson Bay, the Davis Strait and off Amsterdam Is. off northern Spitsbergen; the black right whale might even return in the distant centuries to the waters off St Jean de Luz.

Whalers were captured in wartime, the Seven Years' War, the American War of Independence, the Napoleonic wars and in the American Civil War. In the Arctic fisheries, 735 vessels were recorded as lost from Dutch and English ports and similarly vessels failed to return from the American fisheries. Such figures are well illuminated by the chapter on 'Dangers of the whale fishery' in Starbuck's (1878) history and of course in Herman Melville's *Moby Dick*.

The industrial pelagic fishery

Introduction – the establishment of the Whaling Commission
(Tonnessen and Johnsen 1982)

The industrialization of whaling was achieved by one man, Sven Føyn; in 1873 he patented his steam-driven catcher armed with gun and explosive harpoon. This vessel and others like it were used to pursue the faster whales, blue, fin and sei in the waters off northern Norway. Following a rapid expansion, the first attempts to control whaling were made by the Norwegians within their own industry in the North Atlantic (for a fuller account see Tonnessen and Johnsen 1982).

During the voyages of exploration in the Antarctic seas, whales were seen in large numbers, and expeditions of whalers were mounted from Norway and other countries to exploit them. At this time the whaling companies still needed land stations, and in 1906 there were fourteen whaling leases on South Georgia, South Orkney, South Shetland, and the Falkland Is., all on the Scotia Arc (Figure 63). The companies were Norwegian, British and Argentinian, but the gunners and crews were all Norwegian. In 1906 the governor of the Falkland Is. issued the first Whale Fishery Ordinance which restricted the number of licences and established boundaries within which each licence holder had the right to catch. Perhaps influenced by the decline in North Atlantic whaling, the British Colonial Secretary took steps in 1912 to initiate a whaling conference because of the potential decline in stocks, for the catch of humpbacks in the Antarctic had already been reduced sharply; it never took place because of the onset of the First World War.

During that war, a British Interdepartmental Committee was established to 'facilitate prompt action in regard to the preservation of the whaling industry' (Anon. 1920). They were much influenced by Bruce's report of the destruction of 200000 or 300000 seals in the South Shetland Is. in 1820–1 and 1821–2, after which stocks were very low indeed. Sir Sidney Harmer (Anon. 1920) drew attention to the collapses of stocks of the Atlantic right whale, the Greenland whale, the Pacific gray whale, the Newfoundland rorqual, the Finnmark rorqual and the white whale: he cited a 'period of great plenty (as in the southern humpback) and culminating in total cessation of whaling activity'. He concluded: 'I feel it my duty to express my very strong opinion that . . . the slaughter of whales must be stopped.' The Committee concluded that the industry be controlled by *limiting the number of catchers*, by preventing the capture of calves and nursing mothers, and they also proposed to minimize the waste in processing the carcasses. The latter was really aimed at the possible floating factories which would work away from the licensing control of the Falkland Islands Dependencies Survey. The major immediate outcome of the work of this committee was the establishment of the Discovery Committee, responsible for research on whales and their environment in the Southern Ocean. *RRS Discovery* sailed for Antarctica on 24 September 1925. It was hoped that the future science revealed would have provided the basis for management; the essential work on ageing was started in time to be used by the Whaling Commission, but more extensive work on population dynamics was not carried out by the Discovery Investigations.

In June 1925 the *Lansing* left the Sandefjord in Norway for the South Orkney Is.; she was a floating factory with a stern slipway and vessels like her could act as pelagic bases for the whale catchers. By the end of the 1920s many factories were working at sea and the British Colonial Office had lost control of whaling. Figure 59(*a*) shows a time series of annual catches of whales in the Antarctic, by major species. It will be seen that the catches of humpback collapsed before the First World War; exploitation and then depletion was transferred to blue, fin, sei and minke in that order. Figure 59(*b*) shows catches of baleen whales and of sperm whales in all oceans, from 1910 to 1977. All three figures were taken from Allen (1980).

In 1926, the International Council for the Exploration of the Sea (ICES) established its Whaling Committee under the chairmanship of Professor Johan Hjort, the distinguished Norwegian marine biologist. At the first meeting in 1927, Professor Gryvel reported on the French Projet de Convention, and J. O. Borley from the British Colonial Office gave an account of the Interdepartmental Report of 1920 (Anon. 1920) and of the work of the Discovery Committee. In 1928 the president of ICES, H. G. Maurice (who was also Fisheries Secretary in Britain), expressed to the Whaling Committee his apprehension that the stock of whales might be falling below the level which would ensure continued prosperity; he said that past history had shown that once a stock was seriously diminished, recovery, if possible at all, was exceedingly slow. Nothing was done in ICES because the Economic Committee of the League of Nations had been asked to consider whether an 'international protection of marine fauna could be established'. This Committee studied the French Projet

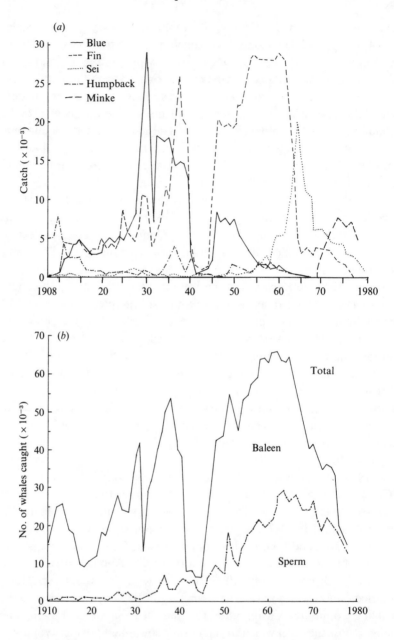

Figure 59. Catches of baleen and sperm whales in the twentieth century: (a) annual catches of five whale species in the Southern Ocean (around the Antarctic and up to the Antarctic convergence) between 1908 and 1977; (b) annual catches of baleen whales and of sperm whales between 1910 and 1980 (Allen 1980).

and a draft Norwegian Bill and recommended the prohibition of the killing of right whales, their calves, suckling, immature animals, and of females with calves of all species; they also proposed that governments should issue licences applicable throughout the world ocean, that the remuneration of gunners and crew should depend on the size of the whale and that statistics should be collected. This document based on the contents of the Norwegian Bill was published in 1931 before much scientific work had been done in Britain or in Norway. These recommendations were taken up by the Geneva Convention of 1931.

At the Geneva Convention, the scientists devised the Blue Whale Unit (BWU), an estimate of quantity weighted by the sizes of different species. But the companies refused to accept governmental control and Hjort and Maurice had to say that voluntary restriction by quotas was the only practical method of control. Britain ratified the Geneva Convention in July 1934, with the addition that the whalers be licensed and that statistics be collected as a condition of the licence (Whaling Industry Regulation Act of 1934); Norway had ratified the Convention earlier and she established the Bureau of International Whaling Statistics. Between 1930 and 1934, there were voluntary BWU quotas between Norway and Great Britain. In 1934, there was conflict between the two countries but a production agreement was achieved in 1935–6.

During the 1930s the science of whale stocks advanced. Mackintosh and Wheeler (1929) published their results on age determination of the blue whale. They showed that the corpora albicantia (remains of the corpora lutea) could be seen in the ovary; if one is laid down regularly, then it is possible to estimate the age of a mature whale. They also plotted the foetal and nursery growth curves, which suggested that blue whales reach a length of 20 m in nineteen months and sexual maturity at twenty-four months. The question of age determination was not settled properly till later, but the apparent demonstration of very rapid growth had one particular effect. Hjort (1933) noted that two stocks of whales, fins off Spain and Portugal and humpbacks in the Antarctic, had both yielded peaks in catch, followed by a collapse. He thought that the Antarctic stocks would not be extinguished because (*a*) the 'enormous cost of whaling expeditions to those distant waters will compel the whaling to stop before the whales are so decimated that they are in danger of extinction, (*b*) the enormous growth rate of whales will favour a comparatively rapid renewal of the stock, *provided the numbers are not severely reduced*' (my italics). Hjort, Jahn and Ottestad (1933) made the first model of a whale population; it differs little in structure from later versions, but was never used. They simulated the decline of the Icelandic stock of fin

whales, but concluded that the time series of catches on Antarctic stocks
was too short. Hjort (1934) noted that the reproductivity of the whale was
very low and so corrected his view of a year earlier; he thought that 'the
technical developments in the whaling industry have everywhere made it
possible for the expeditions to catch more whales than the reproductivity
of the animal permits'. To restrict the number of expeditions 'appears to
necessitate State intervention and, since whaling is carried on in free
international waters, this again presupposes a plan supported by the
Governments of all countries – a condition which seems "utopian".' In
May 1938, the Whaling Committee of ICES (of which Professor Hjort
was still chairman) recommended a reduction of the open season, a
limitation in the number of catchers, an overhead limit of output, a
maximum oil production which no expedition should exceed and special
protection for humpbacks.

In the London agreements of 1938 the capture of humpbacks was
banned south of 40° S, and that of baleens for two years from 8 December
1938 in Area I (around South Georgia), south of 40° S; minimum sizes for
blue, fins and sperms were reduced from those of 1937 agreement, 70 to
65 ft, 55 to 50 ft and 35 to 30 ft (1 ft = 0.305 m) respectively (in order to
placate a new entrant, Japan); the size was less than the mean length at
sexual maturity of females, for example 24 m (79 ft) for blue whales. The
female sperm whale may well be sexually mature at a length of 9 m. In
neither Agreement were the restrictions recommended by the Whaling
Committee of ICES incorporated, even the special protection for humpbacks. The protocol of 1944 established the overall limit of 16000 BWU;
at this meeting was announced the Washington Conference on the Regulation of Whaling. The train of conventions, Geneva, London and
Washington, DC, was established on the conviction (supported by scientific opinion, but not by scientific evidence) that stocks of whales could
not stand the exploitation.

The International Convention for the regulation of whaling met on
2 December 1946 in Washington. Part of the Preamble reads: from

overfishing of one area after another and of one species of whale after another
. . . it is essential to protect all species of whales from further overfishing . . . It is
in the common interest to achieve the optimum level of whale stocks as rapidly as
possible without causing widespread economic and nutritional distress . . . [and]
. . . whaling operations should be confined to those species best able to sustain
exploitation in order to give an interval for recovery to certain species . . . now
depleted.

In the Schedule of the Convention it was forbidden to kill gray whales
or right whales (except for local consumption by aboriginals), calves of

suckling whales and the use of factory ships was forbidden in certain areas. It was forbidden to catch humpbacks in any waters south of 40° S, provided that in each of the seasons, 1949–50 and 1950–1, a maximum of 1250 humpbacks was taken (the peak catch before the First World War was about 8000, see Figure 59(*a*)). It was forbidden to take the baleen whales south of 40° S except between 22 December and 7 April. Minimal sizes were laid down much as indicated above, below the size at sexual maturity. But a whale fishery developed in the North Pacific (Figure 59(*b*).

The provisions of the Schedule of the Convention are amended by adopting regulations

such as are necessary for the conservation, development and optimum utilization of the whale resources . . . based on scientific findings . . . shall not involve *restrictions on the number* or nationality *of factory ships* or land stations, *nor allocate specific quotas to any factory or ship or land station* . . . [and] shall take into consideration the interests of the consumers of whale products and the whaling industry.

These phrases made it a little difficult to execute the control of catching in later years. Presumably the scientists present thought that the overall limit in BWU was adequate. But blue whale catches were already declining (Figure 59(*a*)).

Graham (1943) had already noted that unlimited fishing must become unprofitable. It will be shown below that the quota of 16000 BWU was quite inadequate and in any case the use of the Blue Whale Unit concealed the essential issues. Sir Sidney Harmer and Professor Hjort knew in different terms that the stocks of whales were most vulnerable. The scientists present at the Convention did not; if they had they might have stiffened the phrases on the control of catching.

The period between 1950 and 1964

The International Whaling Commission first met in 1950 and the first full scientific assessment was made by the Committee of Four (it started with three members) in 1964 (see p. 159). During the intervening years the question of age determination of the whales in the catch dominated the Scientific Committee. Ruud (1940, 1945) and Ruud, Jonsgaard and Ottestad (1950) tried to estimate age from the baleen plates; it was later shown by Laws (1961) that this method was only valid for the first five years of life. The major innovation was the use of whale earplugs, conical structures in the outer ear, made of a waxy substance with light and dark laminae (Purves 1955; Laws and Purves 1956). The question was not properly resolved until Laws (1961) examined the annual increment of the corpora in the fin whale and showed that determinations made with earplugs and the corpora albicantia were well correlated. In 1957, Laws presented a paper to the Scientific Committee on the age determinations of fin whales which suggested that the stock was being seriously depleted (by this time the stock of blue whales was well diminished). Slijper, a Dutch scientist, asserted that Purves' method was useless and that the only solution to the problem of declining catches was a large-scale marking experiment (which would have taken a long time (Anon. 1957)). Later, Roe (1967) showed that the earplug laminae are formed in summer and winter. The problem of age determination persisted; no earplugs were ever collected from blue whales (Allen 1980) and the Scientific Committee after 1964 faced the problem of assessment without age determination.

Ottestad (1956) produced a model of the stock of fin whales; he assumed values of juvenile and adult mortalities and magnitudes of annual recruitment; the initial stock was assumed to be 300000–350000 and natural mortality was decreased as stock reduced and the birth rate was increased. Ottestad concluded that under the then levels of capture the stock of fin whales was being significantly reduced. The parameters differ from the later and better estimates, but the model was independent of age determination.

The annual quota in BWU was originally set at 16000. In 1956, it was proposed in the Commission to reduce it to 15000, but there was no agreement because of objections from many nations. In 1957 the quota was reduced to 14500 and the Netherlands Commissioner dissented. In 1958, the reduction in quota was put back for one year. In 1959, the quota was established at 15000 BWU (the Dutch Commissioner had asked for 16000).

In 1959, the Scientific Committee noted that stocks were declining, but they had 'no new recommendation to make though the majority maintain the view they put forward last year, namely, that the balance of evidence indicates an excessive catch and record their regret that, in the last season, the total limit had been raised from 14500 to 15000 units'. In 1960, the Scientific Committee 'are recommending drastic reductions in the catching of all three species, i.e. further protection of blue whales in the Antarctic, total protection of humpbacks of group IV [an area in the Antarctic] and a substantial reduction in the fin whale catch in the Antarctic' (Anon. 1959). The results can be seen in Figure 60, which shows the changes in quota during this period.

In 1961, the Committee of Three was appointed to report within one

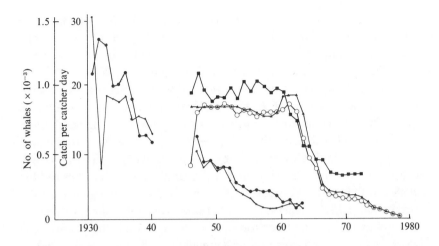

Figure 60. Annual catches of blue whales in thousands in the Antarctic between 1931 and 1963 (●) in thousands; catch per catcher-day of the blue whales in the Antarctic between 1931 and 1963 (●). Catches of all whales in the Antarctic, 1946–72 (○); after 1971, minke and sei whales predominated. Blue whale units per catcher-day (■). Quotas recommended by the International Whaling Commission (▲) in blue whale units; after 1967, the 'quotas' are sums of species quotas. In 1959–62, no quotas were agreed; the numbers represent sums of national quotas. All material is dated to the second year of the Antarctic season (which runs from September to March).

year on the yield that should be taken and on any conservation methods that would increase it; the Three (D. G. Chapman, K. R. Allen and S. J. Holt) were drawn from countries not involved in pelagic whaling. The Commission declared its intention that the Antarctic catch limit should be brought into line with scientific findings not later than 31 July 1964.

The Committee of Four (Gulland had joined the Three) reported in 1964. Mortality rates were estimated from marking experiments executed between 1930 and 1950. Stock densities were estimated from catch per day's whaling and the effort indices were corrected by the tonnages of the catchers. Estimates of recruitment (i.e. the number of young animals joining the adult stock each year) were made. Initial and present stocks were derived with DeLury–Leslie methods (plots of stock density or catch per unit fishing effort on cumulated effort or cumulated catch, respectively; DeLury 1947). The essential theory was simple, $C = F\bar{P}$, where C is catch in numbers, $\bar{P}$ is the average stock in numbers in a given year and F is the instantaneous fishing mortality. The sustainable fishing mortality, $F_S = r - M$, where r is the rate of recruitment (which is density dependent, increasing as stock decreases) and M the instantaneous natural mortality rate. Similarly, the sustainable catch $C_S = \bar{P}(r-M)$. These axioms have formed the basis of whale population dynamics since 1964 (Chapman, Allen and Holt 1964; Chapman *et al.* 1965).

Figure 61 shows the Committee's adaptation of the Schaefer method to the blue whale stock in the Antarctic; it is of course based on an inverse linear dependence of stock density on whaling effort. In the parabola of sustainable yield on stock, the line $\{(r-M) = 0.1\}$ was fitted to the data of sustainable yield at low stock levels; the curve at middle and high stock levels was fitted by eye. The important point is that the catches of blue whales in the period between 1946–7 and 1962–3 were higher than the sustainable yields by a factor of 2 to 3. Table 11 gives the summarized conclusions of the Committee, for four stocks including the pygmy blue whale, which was discovered near Kerguelen by the Japanese in 1960.

The Committee recommended:

(*a*) that the capture of blue whales in the whole region and of humpback whales in areas IV and V (of the southern ocean; see Table 11) be forbidden for a considerable number of years;

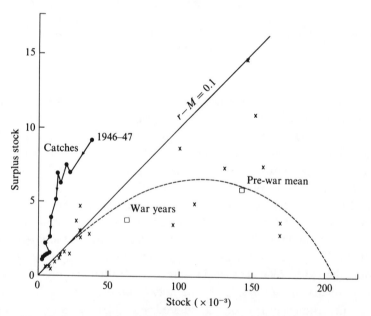

Figure 61. Application of the Schaefer method to the stock of blue whales in the Southern Ocean (see caption to Figure 59(*a*)), surplus production (or catch) on stock; the line fitted to the left of the curve represents a net recruitment (($r - M$) = 0.1), where r is the annual gross recruitment and M is the annual mortality rate (Chapman *et al.* 1964).

Table 11. *Summary of findings of the Committee of Four*

Species	Catch 1961/2	Sustainable yield 1962–3	MSY	Years to reach MSY
Blue	700	<200	6000	50+
Humpback IV, V[a]	810	<100	<1000	50+
Fin	28000	<7000	20,000	5+
'Pygmy blue'	460	—	Few hundreds	—

MSY, maximum sustainable yield.
[a] IV and V denote areas extending from Kerguelen and the Ross Sea, respectively.

(*b*) that the quota of fin whale catches be reduced to 7000 or less (the greater the reduction in the present quota, the more rapidly would fin whale stocks rebuild to the level of maximum productivity);

(*c*) the blue whale unit be replaced by separate quotas for each species.

The Committee of Four also dealt with three other questions in a supplementary report. Slijper's doubts on age determination were answered in three ways: (*a*) Laws' (1961) study validated the use of ear-plugs and the corpora albicantia, (*b*) the Schaefer and DeLury–Leslie methods were independent of age determination, and (*c*) they compared age-length keys from 1930–1 and 1962–3 and found little change in them so the collapse of blue whale catches by a factor of 30 was not due to the ability of older whales to dodge the catchers, as some Commissioners had suggested (Chapman *et al.* 1964, 1965).

Figure 60 shows catches of blue whales from the Antarctic in thousands and the catches per catcher-day. The catches of all whales after the Second World War are also shown, together with BWU per catcher-day. The quotas recommended are also shown, including those in the years 1959–62 when none was agreed within the Commission, although agreement was reached outside it (the figures represent sums of national quotas). The Committee of Four made interim reports and it is obvious that the quotas were reduced after 1961, but all their final recommendations did not become effective until 1967. However, a zero quota for the blue whale did become effective in 1965. This figure records the failure of the Commission until then to save their most valuable resource, the stocks of blue whales. There were three sources of failure – the Commission's structure, the lack of understanding of the scientists in the 1950s and the urgent demand of the companies for oil.

The Commission could amend the Schedule and that amendment then becomes a binding part of the Convention; but nations adhere to the Convention voluntarily and hence are wary of change. The proposals from the Scientific Committee are filtered through the Technical Committee before they reach the Commission, which meant that at this time they were attacked on economic grounds. Any amendment to the Schedule required a majority of three-quarters and there was a ninety-day period for objection; following such objection every other member had potentially a further period of ninety days in which to object. Hence Commissioners were wary of this procedure although it was used by the Dutch to delay proceedings. Decisive and convincing advice might have survived the somewhat rigid procedure, but, if the advice was timid and tempered by economic constraints, the procedure filtered it out.

Because the demand for oil was high, the companies were the villains. Small (1971) showed that Norwegian companies (and by implication, British) would still have made reasonable profits had the capture of blue whales been banned during the middle 1950s. Within the Commission, the companies exerted too much pressure for only an overall quota was in use. However, the ultimate blame must lie with the scientists, who ignored what had happened in the past. Perhaps they were overwhelmed by their lack of exact information and lacked the will to give good advice without it.

Note:

In the verbatim report of the Commission's meeting in 1963, Herrington, the American delegate, said that there were 'several proposals made for limits in the coming season . . . We heard many explanations, but I believe they all boiled down to what is necessary to provide a profitable operation for the entire fleet which operated last year. I do not recall a single consideration based on conservation. There was one proposal, that of the UK, supported by Norway, which was based on the report of the scientists . . . It did not get much much support.' Lienesch, the Dutch delegate, said that 'we have to compare scientific evidence and economic necessity'.

The period between 1964 and 1983

In 1965, the International Union for the Conservation of Nature (IUCN) and the World Wildlife Fund became observers to IWC; they were concerned about the stock of blue whales. In 1972, the UN conference on the Human Environment in Stockholm recommended a ten year moratorium in commercial whaling. When the American delegate to IWC proposed the moratorium, he said that the science was inadequate to protect the stocks. IWC (following its Scientific Committee) rejected this opinion in 1975 saying that there was no overall need for a moratorium; for example, the stocks of sperm whales was not considered to be in danger.

At the Food and Agricultural Organization meeting on marine mammals in 1977 in Bergen, Joanna Gordon Clark (1981) proposed many reasons why whaling should cease. In 1978, an Australian enquiry into whaling asked three questions: (1) do we need whales? to which the answer was 'no', because substitutes were available for all the most valuable uses of sperm oil (but some hungry children might have used the protein); (2) is the method of killing necessary? to which the answer 'no' (anyone reading Sir Alister Hardy's (1967, pp. 183–7) account of whale killing in his book *Great Waters* would be quite appalled); (3) do whales communicate with each other under attack? to which no answer was given. The Australian enquiry proposed that whaling should be banned.

It is not morally wrong to kill whales (for food and other purposes) whilst we kill cattle, sheep and pigs in our slaughterhouses. However, ethically it may be undesirable to kill whales. In the first place, people may have thought it wrong to send the blue whale to join the dodo. Later perhaps they felt, as they looked at the great animals on television, that it was wrong to convert them to margarine. This is the case for the 'conservationists', where conservation means the preservation of the species and not the management of the stocks. Their constituency is a broad one but perhaps a special one confined to parts of the developed world. Their status is really political.

Governments are represented in the Whaling Commission but observers are not. But, following the Stockholm meeting, some governments had assumed the views of conservationists, and these governments pressed for zero quotas. The Japanese and Russians were the main whaling nations, and there were other countries which had never adhered to the Commission; Peru and Chile had one of their own. Hence, there was some danger that if the conservationist case were pressed too hard, the Japanese and Russians might leave the Commission, or that companies would become established in countries outside the Commission.

FAO had been represented on the Scientific Committee since the middle 1960s and IUCN has had an *ad hoc* representation since the late 1970s; but the Committee also invited recognized experts who were paid by conservation bodies. In any international working group of scientists, particularly those with national representation, political factors may be discussed openly. However, scientists can express the consequences of different actions in a common language, and they should give the results of their work as a range of choices, so that Commissioners are free to make their judgements in the political environment.

The history of management in the recent period is one of change. Since

1972, an international observer scheme has come into force, by which observers sail on vessels of nationalities other than their own. The role of the Technical Committee changed, becoming more concerned with regulatory aspects and national attitudes; it remained responsible for enforcement, as it always had been. Instead of meeting irregularly, as it did in the 1950s, the Scientific Committee became overloaded, having to meet two or three times a year. Between 1964 and 1969, stock assessments were carried out under the aegis of FAO. Since June 1969, the Scientific Committee has been responsible for the stock assessment.

In 1965, the Commission agreed that catches of fin and sei whales be progressively reduced so that by the season of 1967–8 they would be below the sustainable yield in that season. Species quotas were introduced for fin and sei whales in the North Pacific in 1969 and in the southern hemisphere in 1972–3. Sperm whale quotas were introduced in the North Pacific in 1970, and in 1971 in the South Pacific; one or two years later the quotas were separated by sex and by area. Since 1931 the right whale, the bowhead and the gray whale had been protected. In 1963, the blue and humpback whales were protected in the southern hemisphere and in 1966 they were protected in the North Pacific.

In 1972, the UN Conference on the Human Environment in Stockholm recommended not only a ten year moratorium on whale capture but also an improvement in management by IWC. The new management procedure was adopted in 1975. Stocks were grouped into three classes: (a) the initial management stock (IMS) one which may be reduced in a controlled manner to that which produced the maximum sustainable yield (MSY), or some other optimum; (b) the sustained management stock (SMS), one from which the MSY (or other optimum) may be taken; (c) protection stock (PS), one in which the stock is below SMS and should be fully protected. The IMS should be 20% above the stock that yields MSY, the SMS should be 90–120% of MSY level, and a stock that was below 90% of MSY level should be protected. If catches had been taken from a stable stock for a long period, it was to be classed as an SMS and quotas were set at recent catch levels, even if stock size cannot yet be estimated. Exploitation of an unexploited stock should not start until a satisfactory estimate of stock has been obtained. By 1978–9, all stocks of right whale, bowhead whale, humpback whale and blue whale were protected. A sustained yield of gray whales could be taken from the North Pacific and similar yields of fin and sei whales could be taken on the North Atlantic. The sperm whale was protected in the North Atlantic and zero quotas were introduced in 1981.

The science in this last phase of the history of IWC developed in two directions, baleen models and sperm whale models. Figure 62 shows the essential structure common to both models; as time passed, the mature stock was distinguished from the exploited stock, the effective minimal size limits were incorporated, the immature component separated from the mature and finally the sexes were treated separately. Estimates of adult natural mortality were made from tagging experiments, changes in age distribution and from the initial age distribution before exploitation (if known). Juvenile mortality was estimated from a balance equation for a stock in equilibrium; given adult mortality, juvenile mortality must be

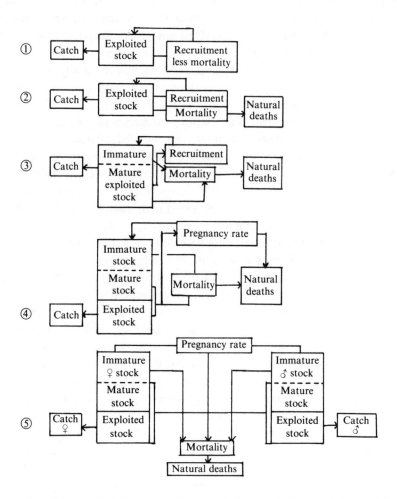

Figure 62. The essential structure of current baleen and sperm whale models (Allen 1980). r, recruitment of young animals; M, mortality of adults.

balanced by the pregnancy rate. If adult mortality is equal to, or greater than, the juvenile mortality, a limiting maximum value can be calculated as function of the pregnancy rate. Recruitment was estimated as the proportion of the current recruited population.

In 1978, the Scientific Committee was asked to examine further improvements in management procedures. Allen (1979) wrote that the Committee had been asked to make recommendations more precise than justified by the accuracy of assessment; some changes in assessment resulted from model changes and not from observed differences in stock. He pointed out that stability could be obtained without a precise definition of MSY. Stocks could be classified as protected or managed stocks from the ratio of current to initial stock (that before the start of modern whaling). The replacement yield (RY) can be estimated from the product of net recruitment rate and current stock; the quota is calculated with the use of factors which account for size of stock and reliability of estimate. This system requires quotas for periods of five years or so and would be assessed by the rate of increase of stock.

In 1979, the management procedure was discussed yet again; for some the science was adequate and for others it suffered shortcomings. The Scientific Committee was asked whether the new management procedure was adequate and whether stock assessments were reliable; they answered that management had improved, but that the sperm whale assessments were weak. They reiterated their advice of the year before, that 'the present Management Procedure does safeguard whale species and secures major stocks for future generations at a very small risk'. They were also asked whether uncertainties can be resolved by moratoria; they could not agree an answer, but noted that observations without capture would be expensive. The Commissioners asked what was the consequence of uncertainty, which implies that they would have preferred certainty. Unfortunately they could not have certainty, at least not in science. But the proceedings devised produced a range of choices.

Allen and Kirkwood (1978) introduced the Baleen model, based on the relationships given above. There were four options: (*a*) age of the mother at first birth is assumed to be density dependent, (*b*) pregnancy rates are available, (*c*) density dependence is driven by the total whale population, (*d*) recruitment is defined as the result of rates of pregnancy and juvenile and adult mortality. The model provides capacity to make the best use of broad and disparate arrays of data (see Figure 62).

In 1978, replacement yields were estimated for the Antarctic stocks and those in the north were classified. The Antarctic is split into six broad statistical areas and in 1980 they were called stock areas. The question of stock identity was thus raised. If a stock suffers from recruitment over-fishing it should be defined biologically, because a gain or loss in recruitment from another stock may bias estimates. It is a very difficult problem and work on it is now under way.

In 1980, the Subcommittee on Minke Whales decided that the adult natural mortality, M, was 0.09; this was based on the relationship between natural mortality and maximum length for a number of species. The pregnancy rate was derived from the balance equation in adult and juvenile mortality (see Figure 62). The stock estimates presented for the Antarctic in that year suggested that an increase of stock had occurred during the period of investigation. Masaki (1976) and Kato (1983) reported a decline in the age of sexual maturity which had started with the year class of 1945; the Baleen model used a time series of age at maturity as the only population response mechanism. Cooke and de la Mare (1983) suggested that there might have been a bias in the age determinations, but a much more important point was that only 15% of the age deter-minations appeared to be valid. At the same meeting, Chapman (1983) suggested that the recruitment rate should be much reduced. In 1983, there was considerable discussion on age determination with no resol-ution and it led to difficulties in considering the biological parameters (Anon. 1984). However, estimates of stock were available from marking experiments and from sightings. When models fail, independent measures of stock are required and it was lucky that the need had been recognized (Horwood *et al.* 1981). By sightings, the Antarctic stock of minke whales amounted to 340 791 and by mark recapture 421 000 (for areas III, IV and V); combining them, a stock of 364 857 was estimated in areas I–V, with coefficients of variation of 18–34%. Thus, the sighting methods appear to be reliable. In considering catch limits there was some discussion of whether the replacement yield (see below) should be 1% or 4% of stock (that is, where the rate of recruitment should be lower than it was earlier thought to be). Thus, a replacement yield of about 3000 animals might have been feasible. Today the stock is not protected but the catch limit is zero. The problem of the Antarctic minke whale is of con-siderable interest, because a yield might have been taken on scientific grounds but, because of conservationist pressure, it was not taken.

The sperm whale models also depend on the basic structure shown in Figure 62, the growth of males and females being separated. Recruitment is expressed in a modified logistic equation and with a function which describes the complex social structure; one male has a harem of ten to fifteen females and, with an even sex ratio, there is a reserve of males. In 1977 and 1978, there was no agreement in the Scientific Committee on how to proceed.

Beddington and Cooke (1981; Cooke and Beddington 1982) introduced a method of assessment based on length distributions applied mainly to the North Pacific stock; with a smoothed length–age key, an assumed equilibrium age structure is updated each year, with the addition of births and the subtraction of deaths. If this model was right the reserve male ratio had fallen below the critical level in about 1966 or 1967. Further, in the model the reduction in numbers from 1979 to 1982 is considerable. However, the model demands a constant selectivity for a particular fleet or area. It had considerable influence, with the consequence that no sperm whales are to be caught, under the rules of the Commission. Work from sightings (Horwood *et al.* 1981; Ohsumi and Yamamura 1981) has suggested that the sperm whale populations are more abundant.

By now, whaling should have ceased. Aboriginals were permitted to take 178 gray whales in 1986 and they probably killed a number of bowheads. Some whales will continue to be caught in the Exclusive Economic Zones of some countries. Since the early 1960s, there were three major scientific achievements, the construction of the Schaefer model for the blue whale stock, the development of the Baleen models and the further development of the sperm whale model. There remains a puzzle about the reproductive rates in those of the gray whale and other species, the stocks of which have been recovered, are of the order used in the models (5–7% year^{-1}), whereas those of stocks such as the bowheads may be much lower. Perhaps there is a critical stock level below which reproductive rate declines, e.g. because of the failure to find a mate.

Conclusion

The early history of stock collapses was well known to Harmer (Anon. 1920) and to J. Hjort; both knew that one way to control whaling might have been to limit the number of catchers. Yet the Commission was forbidden to do that. As Allen (1980) pointed out, the stocks were managed for maximum financial return and some fisheries were extinguished. Indeed the stock of blue whales was reduced to dangerously low levels; Horwood (1986), from sightings, suggests a present figure of 1000–1600 animals. Clark (1973) suggested that the companies should have exploited the whales harder but, in contrast, Gulland (1966) pointed out that the light restraint imposed by the quota of 16000 BWU probably extended the life of the fishery. However, in the 1950s the science was scant and nobody in the field had the authority to persuade the companies to restrain their activities. Then whaling science was transformed by the solution to the blue whale problem proposed by the Committee of Four in 1964. By 1967 most of their proposals had been accepted by the Commission.

From 1972 onwards, the conservationists pressed their case in the Scientific Committee and in the Commission. The new management procedure and later proposals on management were developed as a consequence. Such procedures made it possible to some degree to express the uncertainty which is an essential part of any such prognostication. It was expressed in a range of choices for managers, and such practices are now standard in the fisheries field. The advance in this direction was made first in the International Whaling Commission. The conservationists were quick to exploit the range of choices proffered by advocating the lowest yield in any range. That advocacy should have been limited to the Commission, where there were members who represented governments sympathetic to the cause of conservation. Unhappily there was advocacy within the Scientific Committee, where the range of choices should have made it unnecessary.

Any management structure like the International Whaling Commission is supported by work at sea funded by the member governments. The whaling grounds are huge and lie very far away and as stocks have declined so has the time spent at sea by the scientists. In some centres independent estimates of stock, based on sightings, are being built up. When whaling has ceased, except for those catches made by aboriginals, it would be desirable to follow the recovery of stocks by such methods.

I have limited my account of recent science to the sperm whales and the minke whales because the evidence is fairly firm. But in the same period the Commission has published many papers on other groups of cetaceans, their basic biology and stock characteristics, work of great value independent of the purpose of the Commission, which is now coming to an end.

7

The Pribilov fur seal fishery

The stock of fur seals on Pribilov Is. in the Bering Sea was the first marine population of fish or mammals to be managed within an international convention, which was established in 1911. But the history of the exploitation of the fur seals started in the southern hemisphere; the first sealers reached Tristan da Cunha and the Falkland Is. in 1774. The following account was taken from Riley (1967), Baker, Wilke and Baltzo (1979) and Lander and Kajimura (1982).

Between 1793 and 1807 three and a half million fur seals were taken on the island of Mas á Fuera (now Islo Alejandro Selkirk); by 1807, 'business was scarcely worth following' and by 1824 the animals had abandoned the island (Clark 1887b): 1.2 million seals were taken from South Georgia. Between 1819 and 1822 there was indiscriminate slaughter on the South Shetlands; Weddell (1825) reported that 350000 were killed by forty-six British and forty-five American vessels. Indeed W. S. Bruce told the Interdepartmental Committee (Anon. 1920) in 1920 that 173398 were killed in 1820 alone.

Amongst the islands exploited were Diego Ramirez near Cape Horn, Juan Fernandez (Islo Alejandro Selkirk or Mas á Fuera), Bouvet, Sandwich Land, Desolation (Kerguelen), Heard's Is., Aukland Is., Bounty Is. and Stewart Is., St Felix, St Ambrose, Gough Is. and the Galapagos Is. In the third decade of the nineteenth century, the bonanza was over and the sealers turned for oil to sea elephants, particularly on Prince Edward Is. and Crosier Is. In the first decades of the fishery (up to 1820 and 1830), skins were sold in Canton (Clark 1887b) (Figure 63(a)).

Pribilov discovered the islands named after him in 1786 when he was searching for the fur seals after they had migrated north through the Aleutian passes. In 1799, Tsar Paul gave a charter to the Association of Siberian Fur Merchants and they formed the Russian American Company, with exclusive use and control of all Pacific coasts of America and of all islands in the ocean. In 1822, Governor Mooravyev decreed that young seals be spared, but in 1827 his successor increased the annual catch to 40000. By 1834, the herd had dwindled, catches were reduced to less than 10000 each year and the killing of females was forbidden (but this was not enforced until 1847). In 1841 the kill gradually increased. Figure 63(b) shows the islands in the North Pacific where the seals are taken.

In 1867, the territory of Alaska was bought by the United States and in the following year the capture of fur seals was prohibited. In 1869, the islands were set aside as a special reservation. The Alaska Commerical Company acquired the first of two twenty-year contracts in 1870. Also in that year, an Act was passed to prevent the extermination of fur-bearing animals in Alaska; no seal was to be killed at sea. The first regulations for the conduct of the harvest were issued in 1872: no females or seals under one year old were to be killed. Between June and September, 75000 were to be taken on St Paul and 25000 on St George; some were allowed to be taken outside the season for food for the native inhabitants. In 1874, the proportions within the quota of 100000 were relaxed but the rules were enforced by special agents of the Treasury department.

The Indians on the west coast of North America had always speared the migrating animals. Pelagic sealing started in the early 1860s. There were four major grounds, Farallon (Point Conception to Point Arena in California), Vancouver (between the northern end of Vancouver Is. and the mouth of the Columbia river), Fairweather (Sitka to Middleton Is.) and off the Pribilovs. The vessels used were 70 tonne schooners and by 1894 they took 62000 seals; but many injured animals sank, unrecorded. In 1892 the United States seized and confiscated vessels in the Bering Sea, which led to conflict with Britain. An international arbitration tribunal met in Paris and concluded that the United States had jurisdiction over 3 miles (4.8 km) from the shore. In 1897, American citizens were forbidden to catch seals on the open sea. In 1919, the United States government assumed full charge of the exploitation. The North Pacific Fur Seal Convention was established between Britain (representing Canada), Japan, Russia and the United States in 1911; pelagic sealing was prohibited from 1 May to 31 July within 96 km of the Pribilovs, except by aborigines using primitive weapons. In 1898 Jordan and Clark counted 215900 seals on the islands. The countries shared the skins from the Pribilovs, Robben and Commander Is. Between 1911 and 1917 the seals were only killed for food on the islands. An age–length relationship was established in 1918, from the measurement of seals of known age branded as pups in 1912. Between 1923 and 1932 a minimal yearly breeding reserve of several thousand bachelors was established, each of which was branded or from which a patch or fur was shorn. Between 1932 and 1955 the capture of males was

(a)

(b)

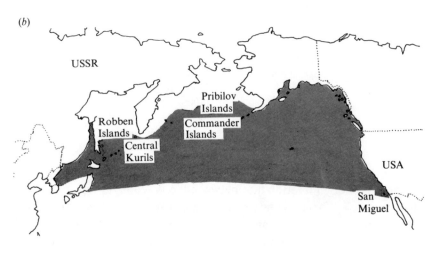

Figure 63 (*a*) and (*b*). Charts of the islands where seals have been exploited.

limited to animals of 104–114 cm in length (about three years of age) from mid June to the end of July. Because the number of harem bulls and killed males did not increase after 1940, it was then thought the maximum stock had been reached. In 1941, Japan withdrew from the Convention and the herd was protected by an interim agreement between Canada and the United States. After the Second World War, Japan ceded Robben Is. and the Kurile Is. to the USSR. In 1958, a new convention was agreed between Canada, Japan, the United states and the USSR.

Figure 64 shows the number of skins taken between 1820 and 1975. There was an initial period of exploitation in 1797–1821, when 1 282 374 seals were killed (about 50 000 year^{-1}), followed by reduced catches in mid century. After the Americans acquired the Pribilovs, catches remained high for twenty years and pelagic sealing reached a peak. Then catches declined for two decades and after the convention was signed in 1911 the stock recovered and so did catches, which remained steady during the 1940s and 1950s, after which problems in management emerged (Riley 1967; Lander and Kajimura 1982).

The northern fur seal breeds on the Pribilovs in the Bering Sea, on Commander Is. off Kamchatka, on Robben Is. off Japan, on the Kurile Is. and on San Miguel Is. off California (since 1968; Figure 63(*b*)). In 1984 the stocks amounted to:

Islands	No. of seals
Pribilovs	871 000
Commander	200 000–225 000
Robben	70 000–80 000
Kurile	45 000–50 000
San Miguel	4000
	1 190 000–1 225 000

Anon. (1984).

There is enough intermixture between the breeding groups to ensure that the whole North Pacific stock is a single stock in the genetic sense, although the groups are managed separately. On Commander Is., 12–21% of the males of three to four years of age originate on the Pribilovs, but only 1% from all Asian rookeries reach the Pribilovs (Lander and Kajimura 1982).

Elliott (1884) described the biology of the Pribilov fur seal. Kipling (1891) adapted his account in his story 'The white seal' in *The Jungle Books*; the link was established by Cushing (1970–1). The harem bulls make for Novastoshnah on St Paul (the oldest arrive first) and 'spend a month fighting . . . for a good place on the rocks as close to the sea as possible.' They stand over a metre clear of the ground on their front flippers and weigh nearly 320 kg. A bull fights 'covered with scars and frightfully gashed; raw, festering and bloody, one eye gouged out'. From Hutchinson's Hill, 'you could look over three and a half miles of ground covered with fighting bulls'. To hold their territories they fast for three months at least.

The cows migrate quickly from the Strait of San Juan de Fuca to the Pribilovs. The bull grabs the cow on landing and the pup is born quite soon afterwards. It is fed once every two days and when it has grown a

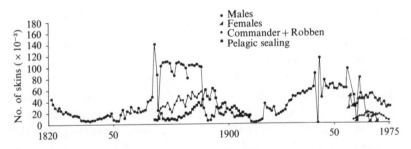

Figure 64. Number of skins taken on the Pribilov Is. between 1820 and 1975.

little it can crawl inland where 'pods' of hundreds or thousands spread in sleep and sport. At the age of six weeks the pup learns to feed; it can swim from the moment of birth (Lander and Kajimura 1982). In late October they put to sea with the adults. The females mature at three years of age and males at four or five; the peak breeding of females occurs between the ages of seven and fourteen and the harem bulls between nine and fifteen. The bulls are active in harems for three or four years only. Mating takes place in June and July, one week after the birth of the pup (Lander 1975).

The adolescent males return a year later and go inland through the rookeries to an area about a kilometre inland. Such animals, the 'holluschickie' of two to five years of age live together in 'solid platoons and obliterate every spear of grass'. It was easy to lead them to the killing ground, by rattling a few bones and stopping every now and then to prevent overheating. The young seals were clubbed and skinned. The quotations above are taken partly from Elliott (1884) and partly from Kipling (1899). Figure 65 reproduces Elliott's drawings of the fishery.

The 'holluschickie' yielded the best skins. A harem comprised about twenty cows and a surplus of males was assumed. The first model of the stock was written by Elliott (1884). About one million pups were born each year; 1% were lost by the time they left in October to November. Only half survived to the following year, the rest having been taken possibly by killer whales, which leaves 225000 males (with a sex ratio of unity). The females live for nine or ten years and the males for fifteen or twenty, so only one male in fifteen is needed; in fact, age distributions were not made until later. Scheffer (1950*b*) used the growth ridges in the cementine of canine teeth for this purpose). If the ratio of one in fifteen is doubled and if every fifth male is set aside, a safe harvest of 180000 could be taken; Elliott (1884) had originally suggested that the 'holluschickie' could be killed to the limit. During the period 1870–90 about 100000 young seals were captured each year. We now know that the death rate of pups at high stock could be as high as 17%, that of the growing males might be as much as 20% year^{-1} and that the production of pups during the high period of the 1950s was never greater than about 450000 (Lander and Kajimura 1982). More detailed estimates of mortality are 50% in the first year, 20% year $^{-1}$ during the second and third years; from ages three to seven the males suffer a mortality of 20% year^{-1} and adult females 11% year^{-1}. The breeding bulls die at the rate of 38% year^{-1} (Lander 1979). Further, the estimate of stock was based on the assumption that each animal needed 2 m^2, whereas in fact they need 6 m^2 (Hanna 1926).

Harry (1974, quoted by Lander and Kajimura 1982), suggests that the

stock in 1867 was 2–2.5 million, in the range of the pre-exploited population and that the pups would have comprised about one quarter of the
stock, about 500 000. In 1911–12 the stock was as low as 300 000. The
reduction was due to overexploitation on the islands and on the open sea.

The first modern model of the stock was written by Kenyon, Scheffer
and Chapman (1954) on the basis of a logistic equation. Chapman's
(1961) model was somewhat more developed. There were six categories
of animal by age and sex, pups, yearlings, bachelors aged two to five,
cows of more than two years of age, idle bulls and harem bulls. The data
comprised bachelor counts (1918–55) and counts of females from 1956,

Figure 65. The exploitation: (*a*) the sea lion hunt; (*b*) top: an oil pouch of sea lion
stomach; and seal meat frame (top left); bidarrah or boat covered with sea lion
skins (top right); sealer's houses(bottom); (*c*) the holluschickie drive to the killing
grounds. All these pictures were drawn by H. W. Elliott (Brown Goode 1887*a*).

age compositions from 1950, tag recoveries from pups from 1947 (except 1950–1), dead pup counts from 1950 (except 1952) and the percentage of pregnant females from the pelagic samples. The year classes 1947–55 were tagged and the estimates of stock (with Chapman's modified Petersen method; see Chapter 9) ranged from 425 800 to 665 000. But Chapman (1964) showed that tagging mortality had biased these estimates. In 1961 Chapman showed that the number of dead pups increased sharply from the low stock period (1914–22, 1–2%) up to 17% in the period of high stock in the 1950s; Shaughnessy and Best (1982) give rather lower figures for the 1960s (6.3–14.5%).

Chapman (1961) assumed that the net gain in weight of the pups per unit time depended upon the time spent feeding by the mother. The animals were further assumed to graze in a circle around the island and the net food intake averaged over this circle was the integral of the net gain in weight per unit distance ($AE^{1/2} - BE$, where E is the number of pups and A and B were constants). Then if survival depends on food, the number surviving, N, $= K(AE^{1/2} - BE)E = \alpha E^{3/2} - \beta E^2$, where K is the fraction of time spent travelling to a given range, and α and β are constants. With this model, the maximum sustainable yield was obtained at a stock of 400 000, not very different from the result obtained with the logistic model of Kenyon *et al.* (1954). But Chapman's model took into account the observed density-dependent mortality of the pups in the first weeks of life on land. The little animals are vulnerable to hookworm (Olsen 1958), a spirochaete, and are prey to starvation. Nagasaki (1961) arrived at similar estimates of the maximum sustainable yield with a logistic and with a Ricker curve (see p. 222).

Chapman (1964) asked, why had stock estimates increased between 1952 and 1956, followed by decline? He noted that stock estimates were biased upwards from tagging experiments because of underestimates of tagging mortality. In 1973, Chapman returned to the problem and set catch as recruitment minus replacement rate, a function of adult mortality; stock for maximal yield was estimated from measures of average recruitment at three levels of stock. Later (Chapman 1981) he suggested that the early decline after 1956 was due to the deliberate harvest of females.

Anon. (1980) describes the problem of management. In 1958, the Interim Convention on Conservation of North Pacific fur seals was established to achieve the maximum productivity which yields the greatest harvest from year to year. As a result of restrained capture of three-year-old males for some decades, there was an 'excess herd'. In order to reduce overcrowding on the rookeries and to improve pup survival, females were taken in quite large numbers between 1953 and 1963, but this practice stopped in 1968. In the late 1960s and early 1970s the stock of adult males declined sharply. This trend, however, was reversed between 1973 and 1976. In 1973, the island of St George was closed to exploitation in order for a comparison to be made with the exploited stock on the island of St Paul.

Eberhardt (1981) made a thirty to fifty year simulation of the fur sea stock, with a low rate of population increase ($r = 0.06$) and reasonable survival rates. The model stock was perturbed and it returned to the initial conditions, but after a rather long time, as might be expected from low vital parameters. The conclusion is that rates observed over relatively short periods may be biased by earlier events, for example in management procedure. An analysis of the possible observed vital rates by Smith and Polacheck (1981) suggests that such rates might change slowly with time; they also believed that the population is regulated by very small changes in more than one vital parameter, as distinct from the simple and attractive mechanism proposed by Chapman (1961).

Chapman (1981) wrote that 'it has never been possible to count all elements of the population'. For example, critical data on the growth of the herd between 1924 and 1947 are missing. The estimates of pup numbers in the 1950s were distorted, so the results from the models are biased. During the period of recovery after 1911 there were three management strategies: (a) spare a breeding reserve of subadult males; (b) harvest more immature males, and (c) harvest females. Because of the slow response of the population the last two were perhaps confounded. Chapman listed four possible sources of difficulty: (1) The age structure may have been modified by management manipulation. (2) Lags may have delayed the expected response. (3) The fishery for Alaska pollack may have affected the feeding of the lactating females. Scheffer (1950a) showed that pollack was a predominant food item in the Bering Sea; Taylor, Fujinaga and Wilke (1955) noted that the same was true off Washington State. As the western population from the west coast of America and the Pribilovs migrate through the Unimak Pass into the Bering Sea (Fiscus, Baines and Wilke 1964), pollack must have been an important item of diet. (4) Fur seals may have been lost in the Japanese salmon drift nets; as many as 3500 are taken each year (Lander 1979). They may well suffer from entanglement in other forms of synthetic gear.

The capture of females was proposed during discussions in 1954 and 1955 on the Convention which was signed in 1958. It was assumed that the reduction in the stock of females would be compensated for by increased pregnancy rate. The production of pups started to decline in the late 1950s, but it may well have stabilized by the early 1960s (Kenyon *et al.* 1954; Lander and Kajimura 1892); this was when the relatively high catches of females were made. The number of harem bulls decreased from the early 1960s onwards (Figure 66; Lander and Kajimura 1982). Since the start of the new Fur Seal Commission based on the 1958 Convention, research has been carried out on the fur seal stocks on the Pribilovs and elsewhere.

For a long time the recovery of the Pribilov herd since 1911 was considered a model of management practice. In the sense that the stock did recover, the judgement stands. However, since the middle 1950s, the

The provident sea

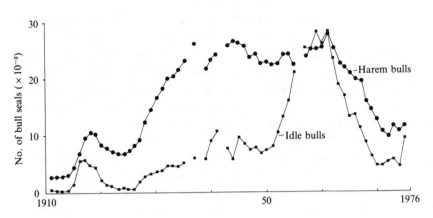

Figure 66. The numbers of idle bulls and of harem bulls, 1912–76.

stock has declined, perhaps initially because of the management procedures, particularly the decision to kill female seals. But the later decline of stock may well be caused by the entanglement of seals in nets of synthetic fibre (which do not decay). The scientific problem was the low rate of natural increase in the population. Consequently the perturbation generated by the overkill of females became apparent rather slowly and recovery was slower than anticipated. In the future there will be considerable need to estimate the vital rates of the population.

8

The Newfoundland harp seal fishery

There are three stocks of harp seal in the White Sea, on Jan Mayen and off Newfoundland. One group leaves the east side of Baffin Bay in November and migrates southward to the east coast of Newfoundland forming the Newfoundland stock. The sea around Newfoundland freezes in late February and the females whelp in groups on the ice; they need ice about 5 cm thick on return from the feeding ground to the whelping ground. The pups lose their valuable white fur after they are deserted by their mothers at 25–27 days, when they lose weight and learn to swim (Mosdell 1923; Lavigne 1981). The adult harp seals are 1.8–2.0 m in length and weigh 100 to 150 kg. The females mature at five years of age and the males at six. The pregnancy rate is greater than 90% and Sergeant (1975) drew attention to possible density-dependent changes in maturity. The pups are born in early March, and are followed by mating; gestation takes 11½ months. The animals live for about thirty years. The pups are killed with clubs or *hakapiks* (pickaxes) and adults with rifles. The pups are used for oil and fur and the adults for leather and meat (for the Canadian landsmen).

From early records, seals were taken sporadically before the end of the eighteenth century (Prowse 1896). The modern fishery started in 1792, when 7000 seals were taken off Bonavista in Newfoundland, of which two thirds were harps and one third hooded. The hooded seals come from East Greenland and their pups are whelped on rough hummocky ice north-east of the harp seal patches. In 1795, 4900 seals were taken by nets off Conception Bay and Labrador; each boat fished twenty nets and was crewed by four or five men. The seals caught were immature, or bedlamers (*bêtes de la mer*). In 1796, four schooners were fitted out in St John's and several from Conception Bay; the vessels displaced 40–75 tonnes and carried thirteen to eighteen men. In 1862, wooden steamers (built in Dundee) were introduced and replaced the schooners fairly quickly. Aeroplanes were used in 1922 for scouting and even for sealing. Between 1863 and 1917, forty-two steam vessels were lost (Mosdell 1923).

Figure 67(*a*) shows the catches from the early nineteenth century until 1970; it gives the average catches of seals (which includes a minority of hooded seals) between 1818 and 1920 (Mosdell 1923); Figure 67(*b*) shows the catches of pups ('young seals on the ice' according to Mosdell), bedlamers and old harp seals from 1895 to 1920 (Mosdell 1923); Figure 67(*c*) illustrates the catches of pups, bedlamers and old seals between

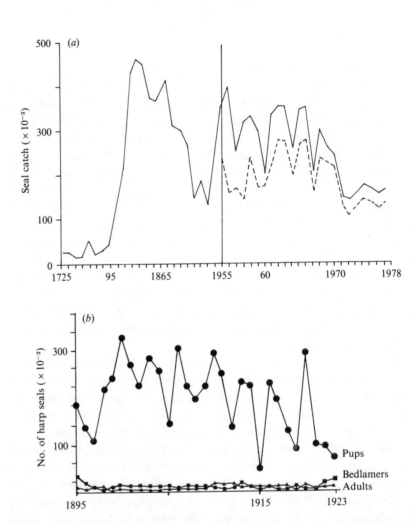

Figure 67. Catches of harp seals from the early eighteenth century until 1977; (*a*) average catches between 1725 and 1977 (Mosdell 1923; ICNAF statistics by decades until 1955 and after that year by year); (*b*) catches of pups, bedlamers and adult harp seals 1895–1923 (Mosdell 1923); (*c*) catches of pups, bedlamers and adult harp seals between 1938 and 1971; (*d*) distribution of sealing grounds.

1938 and 1971. Figure 67(*d*) shows the main regions of capture. Between 1971 and 1976, the average catches of pups amounted to 166366. Because Norwegians took part in the capture of seals since 1938, the stock was managed with scientific advice (The International Commission for North West Atlantic Fisheries, ICNAF, from 1966 and since 1977 from the North Atlantic Fisheries Organization, NAFO). For a decade or so,

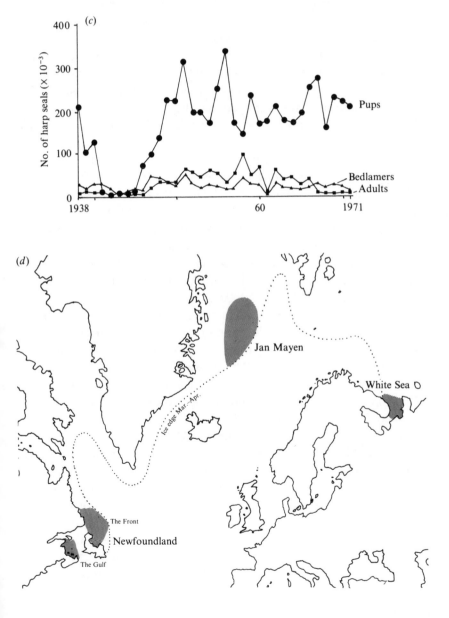

much concern as to the destruction of the pups has been expressed by the conservationists, probably because the killing has been shown on television throughout the world.

Because of their particular way of life, the direct enumeration of stock is a little difficult. Sergeant (1975) lists four methods: (*a*) photographic surveys of whelping adults from the air, (*b*) recapture of tagged pups, (*c*) the sum of catches of pups if none survived, and (*d*) by plotting catches on the relative magnitude of year classes. The photographic survey has some limitations; navigation is difficult over the pack ice, the pups are too small and white and the mother seals are often off the ice foraging for food. However, Lavigne *et al.* (1975) have shown that the pups can be detected well with ultraviolet photography. The tagging methods should work well but the tags must be mixed properly into the population. The use of the four methods suggested that the abundance of pups declined from about 400 000 in the 1950s to about 200 000 in the 1970s (Sergeant 1975).

Allen (1975) gave the distribution of pregnancy by age. He also showed that the fishing mortality rate amounted to 7% in the late 1950s, when the total mortality rate was 15%; in the early 1970s the rates were 5% and 12%, respectively. Hence the rate of natural mortality was about 7–8%. From pup catches and from the ratio of pups to adults, stock estimates were derived. Life tables were used to estimate pup production and the data fitted by linear and Ricker stock/recruitment curves. From a reconstruction of the stock of adults by age, Allen found that it declined from about a million in 1951–4 to 650 000 in 1971.

In 1966, the ICNAF noticed that the stock had been declining for fifteen years. In 1968, the duration of the hunt was shortened and the killing of adults in whelping patches was prohibited; airborne sealing was banned in 1970. A quota of 245 000 was established in 1971, but the catches did not reach the quota. In 1972, a total allowable catch (TAC) of 150 000 was agreed in ICNAF and sealing in the Gulf of St Lawrence was banned; in 1976, the TAC was 127 000 and 160 000 were killed. A quota of 180 000 was agreed in 1977 (Bonner 1982).

During this period a number of models of the harp seal stock were written. Capstick, Lavigne and Ronald (1976) developed Allen's earlier model with a density-dependent mean age at maturity of female harp seals; with the 1976 catches they would expect the stock to collapse in the 1990s or later. Lett and Benjaminsen (1977) made a simulation model based on an age structure derived from sex ratio, pregnancy rate, increase in maturity with age and natural mortality. The age of maturation and the pregnancy rate were density dependent. A maximal stock was established and it was concluded that the stock was recovering, under the regime of TACs. Winters (1978) and Lett, Mohn and Gray (1981) developed models in sequential population analysis; that is, cohort analysis started in the latest year by stock quantities rather than by fishing mortality (see Chapter 14). The age distributions were based on age determinations from jaw samples; but Doubleday and Bowen (1980) showed that the age determinations of older animals were variable. Natural mortality was estimated from the 'survival index', the ratio of numbers of pups minus catch to numbers of the same year class in later age samples. Three possible density-dependent factors were identified (natural mortality, mean age of whelping, proportion of females whelping). A maximal catch of 200 000 was deduced at a stock level of 1.3 million.

There was some criticism of these models because the age structure depends on the estimate of natural mortality derived from 'survival index' (the hunting mortality of adults is very low). Beddington and Williams (1980) developed an age model from the data on catch at age; pups, catch and stock with natural mortality generate pups again; an arbitrary density-dependent function was used. They calculated the cumulative χ^2 for the observed catch at specified age and that expected from the model at minimum χ^2, $M = 0.1375$, rather higher than that of the Canadian models (about 0.10). Bowen and Sergeant (1982) examined the mark-recapture experiments and concluded that in 1978–80 the pup population amounted to about half a million; the hunting mortality is low and hence the estimate of tags recovered must be variable. Goodman (1982) pointed out that the highest estimate of natural mortality implies that the stock is declining and the lowest that it is increasing. This is the crux of the argument between the Canadian government (which used the models of Lett and Winters) and the conservationists (who used a number of models, including that of Beddington and Williams). Goodman also suggested that considerable resources be devoted to a large tagging experiment and an extension of the ultraviolet photographic survey. How nice to hear a distinguished expert on animal populations recommend more work at sea. The future of the harp seal fishery is uncertain, since the European Community has banned imports of sealskin.

9

The origin of fisheries science

Fisheries science has two roots, in the development of marine biology in the nineteenth century and in the first industrialization, described in Chapter 5. The history of marine biology is outside the scope of this book but is indicated in the Victorian interest in expeditions at sea and in the establishment of marine laboratories.

Marine biology

The Victorian interest in marine biology

The Victorian age was one of discovery and one of the first explorers in marine biology was Edward Forbes. He was born on the Isle of Man in 1815 and was first an artist, then a doctor and finally a marine biologist. He was primarily a student of molluscs and echinoderms, both present day and fossil. He dredged in the Irish Sea, off the Shetlands, in the Firth of Forth, off Algiers and in the Aegean. He distinguished provinces in the sea and described the zonation of animals in depth. He said that the deep ocean was 'an abyss where life is either extinguished or exhibits but a few sparks to mark its lingering presence' (Forbes 1849). He illustrated his monograph on British starfishes with vignettes (Figure 68(*a*)) and self portraits (Figure 68(*b*)) and commemorated his exploits in his 'Dredging Songs' (Forbes 1841). His contemporaries were the great collectors and systematists of the age, but in Forbes' life there is a foretaste of the expeditions and of the scientists who went to sea.

The Victorians were avid for the information then becoming available. It was purveyed by many popularizers, of whom Gosse was the most prominent. He started his working life as a clerk in a seal fishery office in Newfoundland. He became a serious zoologist and wrote two notable systematic works on sea anemones and corals (Gosse 1860) and on rotifers. In mid century he wrote a number of books illustrating animals and popularizing their lives. Figure 68(*c*) shows one of them (Gosse 1856), a

Figure 68. The view of the Victorian naturalists: (*a*) view of Leith roads (Forbes 1841; (*b*) vignette from Forbes (1841); (*c*) dredging off Whitenose (Gosse 1856).

Figure 69. Victorian holidays by the seaside: (*a*) seaside walks (Houghton 1870); (*b*) seaside holidays (Anon. undated); (*c*) sandcastles (Anon. undated).

naturalist working a dredge to take specimens from the sea bed. At this time the Victorians spent their holidays by the sea (Figure 69) at places like Tenby or the Devonshire coast and Gosse's books led them to particular places to find the animals which he had illustrated so beautifully.

The expeditions (Murray and Hjort 1912)

The great expeditions in the ocean started with the survey voyages of James Cook in the *Endeavour* and the *Resolution*; Cook was a naval officer, the vessels were provided by the Navy and the voyages were sponsored by the Royal Society of London. Such links between science and naval authorities were common in most countries in the nineteenth century. Charles Darwin sailed with Captain FitzRoy in the *Beagle* (1831–6) where he collected the information later transmuted into the *Origin of species* (1859). In 1818, Captain Sir John Ross sailed in the *Isabella* to Baffin's Bay, and between 1839 and 1843 *Erebus* and *Terror* explored the Antarctic under the command of Captain Sir James Clark Ross, whose chief scientist was Joseph Hooker, a collector of diatoms. These two ships were lost later when Sir John Franklin searched for the North West Passage. Some of these voyages were largely exploratory in the geographical sense.

In 1838 and 1840 the United States Exploring Expedition sailed in the *Vincennes* and the *Porpoise* under the command of Captain Wilkes; J. D. Dana was the chief scientist. In 1842 Edward Forbes visited the Aegean in the *Beacon* and laid the foundations of his studies of depth zonation in the sea. In 1850 and subsequently, Michael Sars collected animals in deep water off the coast of Norway. T. H. Huxley put to sea in *Cyclops*; on this voyage he found *Bathybius*, which was later found to be a precipitate from sea water to which alcohol had been added. In 1860, in the *Bulldog*, Wallich found thirteen starfish on a sounding line in 2520 m. In 1867 Louis Agassiz and the Comte de Pourtalès worked from the *Bibb*, a vessel of the US Coastal Survey. In 1868, Wyville Thompson and W. B. Carpenter dredged in deep water west of the Faroe Is. in the *Lightning*. In 1869 Wyville Thompson and W. L. Carpenter worked west of Shetland and off Rockall in the *Porcupine*. Up until this time most of the ships had collected samples from the sea bed in fairly deep water. Animals new to science were bottled, labelled and identified; many beautiful drawings were made of them by the scientists who had described them.

In 1872, Wyville Thomson sailed on the *Challenger* and circumnavigated the globe in scientific exploration under the command of Captain Nares. The science of oceanography started with this expedition. The depth and character of the seas and the sea water were examined across the oceans and in remote places. The record remains in the reports edited by Sir John Murray, a scientist aboard.

In 1880, Alphonse Milne-Edwards started a series of cruises in the *Travailleur* and the *Talisman* to the Bay of Biscay, the Mediterranean and the Canaries. From 1887 to 1890, Alexander Agassiz explored the Caribbean for the US Coast and Geodetic Survey in the *Blake*. Through the 1890s he also worked for the US Fish Commission in the *Albatross* on the continental shelf off the eastern seaboard of the United States. From 1885 onwards Prince Albert I of Monaco conducted a series of cruises in the North Atlantic in *Hirondelle*, *Princesse Alice I*, *Princesse Alice II* and *Hirondelle II*; he released drift bottles in the North Atlantic and made many oceanographic observations there. Admiral Makarov made a long expedition in the *Vitiaz* from 1886 to 1889. Victor Hensen sailed in the *National* in the North and South Atlantic in 1889 and the results of the Plankton Expedition were the first record of plants and animals from the midwater on an oceanic scale. In 1895–6, the Danish *Ingolf* expedition worked in the Atlantic; in 1897–9 the Belgian vessel *Belgica* sailed to the Antarctic, and in 1898–9 the German *Valdivia* expedition visited the Atlantic and Indian Ocean under C. Chun.

Before the *Challenger*, the vessels often had a general exploratory purpose, surveying coast or sea bed. After Forbes' discovery of depth zonation and his thesis that the sea below 800 m was azoic, there was an impetus to the exploration of the sea bed and the animals that live there. After the *Challenger*, the expeditions became more numerous and specialized as more European nations mounted them. Exploration continued in the twentieth century on an expedition scale, but fisheries science had already started in the last two decades of the nineteenth century.

Marine laboratories (Murray and Hjort 1912)

The collections made by the expeditions were examined in the universities and at first there was little need for special marine laboratories. The University of Paris established one at Concarneau in 1859 and another was built in Arcachon in 1863. Much of the credit for the general establishment of marine laboratories must go to Anton Dohrn: he read a paper at the British Association in 1870 which proposed that marine laboratories be built. In 1872, he built the Stazione Zoologica in Naples with his own funds. He established 'tables' for visiting scientists, which were working spaces with facilities; the specimens were animals taken from the Bay of Naples.

In 1883, the International Fisheries Exhibition was held in London. It was a large exhibition with a considerable scientific input introduced by a notable address by Professor Huxley, then President of the Royal Society (for a fuller account, see Chapter 5). A Royal Society Committee recommended the foundation of the Marine Biological Association in 1884 and its laboratory in Plymouth was built in 1888. In 1884, the Scottish Marine Biological Association was founded and scientists worked from its barge, the *Ark*; in 1885, a laboratory was built at Millport. In the same year, at Wood's Hole in Massachusetts, the US Fish Commission built a laboratory and the Marine Biological Laboratory was established in 1888. In 1889, Petersen worked the Danish *Station*, a barge which could be towed from place to place. The Gatty Laboratory at St Andrews started work in 1892, as did the Liverpool University laboratory at Port Erin in the Isle of Man. Hjort set up a laboratory at Drøbak on the Oslofjord in 1897. The Scottish Fishery Board came into being in 1882 and the Marine Laboratory was built in Aberdeen in 1898. Prince Albert constructed his Monaco laboratory in 1899.

This is a long list of famous laboratories all established within two to three decades. Many equally famous laboratories appeared in the twentieth century. Fisheries science originated in the last decades of the nineteenth century, nourished in the scientific climate of the Victorian age. It is interesting to notice that Gosse published his first popularization in the early 1840s when Edward Forbes was still working at sea. The early expeditions of Darwin and the Rosses had returned. Then Forbes' proposal that the deep sea was azoic generated the middle period of *Bulldog*, *Lightning* and *Porcupine*. The *Challenger* inaugurated the last period, that of the great collections by expeditions from many nations. The laboratories were not established until the last three decades of the century, indeed after the industrialization of fisheries had started.

Fisheries science

In Chapter 5, it was shown that as capture was mechanized stocks became depleted. As fishermen experienced the decline in stock density, they worked further afield and started the exploration which has continued well into the present century. The demersal stocks in the North Sea were explored in this way, as were the stocks of Pacific halibut on the shores of the Alaska gyral.

Marine science flourished in the nineteenth century and it was quite natural that scientists from T. H. Huxley onwards should involve themselves in the concerns of fishermen. At the same time governments became persuaded of the need for action between nations. The result was the establishment of the International Council for the Exploration of the Sea in 1902 and a treaty between the United States and Canada to manage the stocks of Pacific halibut was signed in 1923. The development of these institutions in the first two decades of the present century will be described with the science in the hands of C. G. J. Petersen, W. Garstang, W. F. Thompson, E. S. Russell and M. Graham.

Petersen

Petersen (1894) first examined the nature of the stocks of plaice in the Kattegat. There were differences in growth rate and in the number of finrays between the plaice in the Kattegat and those in the Belt Seas and the Baltic. Petersen thought that a stock 'propagates generation after generation with the peculiar stamp of the race'. He tagged fish (Petersen 1896) with two numbered buttons of bone fastened behind the dorsal fin with silver wire; fish released in spring at a length of 18–25 cm were recovered in October and November at lengths of 33–36 cm. Hence, modes in the length could be aged, if a little roughly: the Petersen method. The fishermen told him that twenty years earlier at the start of an intenser fishery there were some large, lean, unpalatable plaice, the *Hanser* or *Praesteflyndere*, like the 'elephant's lugs' recorded by Holt (1893–5) on the Dogger Bank in the North Sea.

The tagged fish did not leave the Kattegat, so Petersen said the stock was stationary. Then the loss of older fish must have been due to fishing. He wrote that fishing power had increased by a factor of 20 or 30 in the twenty years of the developed fishery; 'if we fish the plaice while they are small, we do not get so great a profit'. The fleet should take 'exactly so much as the stock could reproduce by new growth' (of individuals and by recruitment). Natural mortality was low because fish were not obviously diseased and were not found in the guts of the only potential predators, cod and marine mammals. Of fish that reached 25 cm in length, a large proportion should have reached 36 cm if protected and would have produced 'a much greater quantity of meat every year'. This was his growth theory in contradistinction to 'other propagation theory' (we now contrast growth overfishing with recruitment overfishing). Holt (1893–5) recorded that 83% of the North Sea catch of plaice was immature or below the 'biological size limit'; Fulton (1892) had put forward the same view. Petersen thought that the problem of growth overfishing would be solved by the fishermen themselves once they saw the economic advantage of catching the larger fish and allowing the smaller ones to survive

Table 12. *The decline in stock density in the North Sea*

Year	Annual catch (tonnes)[a]	'Smack units'[a]	Tonnes per smack unit
1889	173180	2859	60.6
1890	172055	3086	55.8
1891	180054	3711	48.5
1892	187512	4057	46.2
1893	200281	4307	46.5
1894	215408	4599	46.8
1895	228180	4918	46.4
1896	232034	5620	41.3
1897	225864	6099	37.0
1898	230656	7147	32.3

[a] From Garstang (1900).

and grow. On recruitment overfishing: 'I will not deny that it is possible to fish up a species to such a degree that there are not left individuals enough for the breeding of them, but I believe that we are far from that point with the plaice', a comment which may still be true for that species.

Garstang

Garstang (1900) worked at the Marine Biological Association's Laboratory at Plymouth and introduced two important steps. First, he used the average annual catch per unit of effort as an index of stock; for example, as shown in Chapter 5, the average catch of plaice made by four Grimsby smacks declined by more than a factor of 2 between 1875 and 1892. Garstang's second step was to devise a rough index of fishing power in that one steamer was considered equivalent to four smacks, and he derived an estimate of fishing effort, smack units. He showed that between 1889 and 1898 fishing power in the Grimsby fleet more than doubled and that the catch per unit of fishing effort (or stock density) declined by more than a quarter. He extended the argument to statistics from the whole of the east coast of England, as shown in Table 12.

Later this relationship, the dependence of stock density (as tonnes per unit effort) upon fishing effort (or time spent fishing), became of importance in the hands of W. F. Thompson (see below). The stock density of trawled fish was reduced to nearly half in the last decade of the nineteenth century, as steam trawlers began to dominate fishing.

The International Council for the Exploration of the Sea before the First World War (Anon. 1902)

On the instigation of the Swedish Hydrographical Commission, the Swedish Government, in June 1899, called the International Conference for the Exploration of the Sea, in Stockholm. The conference decided, 'considering that a rational exploration of the sea should rest as far as possible on scientific enquiry, . . . to improve fisheries through international agreements, [and] . . . to recommend investigations . . . for a period of at least five years.' Extended hydrographical observations were to be made each quarter and the biological work was concerned with the distribution of eggs and larvae of marine economic fishes, their life histories, races, migrations, food and enemies. Experimental fishings were recommended, as was a uniform collection of statistics based on that of the Scottish Fishery Board. Petersen's scheme for the recording of uniform trawling operations was put forward, and Garstang specified the overfishing problem, in particular the destruction of immature flatfish (Anon. 1902).

The International Council for the Exploration of the Sea was founded on 22 July 1902 in Copenhagen under the presidency of Professor Herwig, who represented Germany. Three Committees were established on migration, overfishing and Baltic fishes. Garstang distinguished between the general problem of overfishing and the special one of the destruction of immature fish off the coasts of Holland, Germany and Denmark. The central bureau was sited in Copenhagen, linked to the Danish Foreign Office for possible international action. Its first job was to organize the joint cruises; in August 1902 the ships sailed: *Thor* (Denmark), *Poseidon* (Germany), *Huxley* (England), *Wodan* (Sweden), *Michael Sars* (Norway), *Goldseeker* (Scotland) and *Andrei Perwoswammy* (Russia) (*Rapp. Procès-Verb. Cons. Int. Explor Mer*, vol. I).

In February 1904 the overfishing committee considered the trawling experiments; in Garstang's words, 'differences were of a serious character, since they not only precluded the possibility of any reliable combination of the trawling records . . . but would impose great difficulties'. The sampling power of the fleet is very much greater than that of a few research vessels, but that was not obvious until later. The dismay of the scientists was reflected in the fact that the name of the Overfishing Committee was changed to that 'investigating the biology of the Pleuronectidae and other trawl-caught fish'. Indeed Petersen said that 'overfishing was not the most essential question for practical purposes' and that the 'Committee should examine the transplantation of plaice' (Anon. 1904).

Between 1905 and 1914, the plaice question dominated discussion. Petersen, Garstang and Kyle (Anon. 1907) said that 'very large quantities of small plaice landed in the Netherlands and England involve loss to fishermen . . . but until it can be proved that the fisheries are deteriorating [there is] no justification for control'. Garstang (1909) showed that size and density were inversely related, and that the small plaice lived on the Dutch, German and Danish coasts and the bigger fish lived in deeper water. Heincke (1913) showed that the number of small plaice *discarded* exceeded the number caught in coastal waters in summer by a factor of 2–6; from tagging experiments, he estimated the annual fishing mortality at 20–50%. He wrote: 'this useless destruction of millions of under-sized plaice and its possible prevention must be considered as the real root of the plaice question'. In his 1913 report, Heincke said that the decrease in numbers of large plaice was the direct result of intense fishing; he proposed that undersized fish be protected with a size limit of 25–26 cm. No action was taken internationally because of the outbreak of the First World War.

However, in its early days the Council failed. There were two reasons. The first was statistical: all the distinguished scientists were dismayed at the variability of the research vessel catches. Heincke (1913) worried whether a sample was representative and how to manage the difference in English length measurements (in 5 cm groups) and German (in 1 cm groups). Such concerns were proper but the statistics of small samples had not then been developed. The second source of failure was theoretical: Garstang and Petersen diagnosed overfishing, but there was no theoretical basis for resolution of the problem.

Statistics in the early stages

Statistics of the fisheries have been collected for many purposes. One of the earlier complete sets is that on the British white herring fishery from 1808 to 1875, as noted in Chapter 4. Statistics on catches in weight by species from areas in the sea or by ports were collected in most European countries by the last decades of the nineteenth century. Fulton (1908) described the system started by the Scottish Fishery Board; the position of fishing, the amount of fishing, the depth of water and the proportions of large, medium and small fish in boxes were recorded. The statistics were expressed in cwt/100 h fishing in statistical 'squares' of 1° latitude by 2° longitude in each month (1 cwt = 50.8 kg). This system, with minor modifications, was adopted by d'Arcy Thompson (1909), in the International Council, as an example for the collection of statistics. Russell (1914, 1922) described in detail the length measurement made on fish in

English markets, started during the International Council investigations. Probably the most influential work at the time was that of Kyle (1905), the Council's statistician. He attacked the idea that 'statistics of the average catch per boat were alone sufficient to prove the existence of over-fishing and coming exhaustion of the fisheries'. Such an index depended on the total catch and the amount of fishing; Kyle thought that the catch per unit of effort might be reduced merely because there were more vessels in the area, independently of any change in stock density. But of course any use today of this index presupposes adequate sampling. Only a sharp decline in total catch was admitted by Kyle to indicate overfishing. He also disputed the fact that the decrease in size of fish was a sign of over-fishing. Add to this the dismay expressed in the Overfishing Committee at the variability of research vessel catches, and it is not surprising that the earlier results of Petersen and Garstang were not pursued.

Thompson and the Pacific halibut (Thompson and Freeman 1930)

The history of the Pacific halibut fishery as given in Chapter 5 is here recalled briefly. When the transcontinental railways were completed, men and boats were brought from the east coast, as the Atlantic stock became scarcer and pricier. Schooners and small steamers from Seattle worked in Puget Sound in the Strait of San Juan de Fuca and off Cape Flattery. By 1903 boats from Vancouver fished off Queen Charlotte Is. and between 1904 and 1909 off Hecate Strait. By 1899 the first landings were made in Alaska and all the shoreline of the Gulf of Alaska had been explored in 1911.

Exploration was generated, as always, by depletion of the local grounds. By 1909, the stock had been reduced in Puget Sound and in Hecate Strait and the boats spread north-west into deeper water. During the period of depletion near home and exploration further afield, the fleet became more efficient. In 1895, dories were used from small steamers and in 1901, slings were introduced for quicker unloading and nets doubled the rate of landing. Between 1895 and 1910, crushed ice and frozen bait were used at sea and cold storage plants were built ashore. The dories were abandoned in 1913, and replaced by long lines arranged as 9–11 hooks (a skate). The industrialization of the halibut fishery followed the same course as the North Sea trawl fishery as it depended upon railways, the use of ice and the improvement of gear at sea.

The scientific study of the Pacific halibut started with W. F. Thompson (1916, 1917a), when he worked at Victoria in British Columbia. He found differences in weight for age and in the morphometrics of the head, between the stock in Hecate Strait (Area 2 off British Columbia and Washington State) and that off Kodiak Is. (Area 3 off Alaska, see Figure 70). He established that the fish spawn in midwater and mature between eight and seventeen years of age. Off Alaska, the length of voyages had increased by a factor of 3 and the time lost in winter by bad weather had doubled with respect to the stock in Area 2; in this area, stock density declined by a factor of 4 in ten years. Thompson (1917*b*) proposed: (1) to stop fishing in large areas (for example Hecate Strait), (2) to increase the closed season during spawning (in May 1914 the *Pacific Fisherman* had proposed it to protect 'fish which should have been allowed to spawn'), and (3) to limit the number of boats and men employed. Thompson saw that the decline in stock density between 1907 and 1913 was due to fishing and that the cure for depletion was to restrict the fishing effort (or time spent fishing). In this paper he pointed out that the spawner killed in summertime was just as much a loss as one killed on the spawning ground.

The science of the new Commission was quickly established. Halibut were tagged in summertime in both areas (Thompson and Herrington

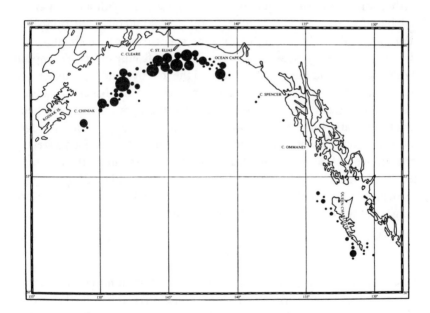

Figure 70. Egg surveys made by Thompson and van Cleve (1936), which show the distinct spawning areas, off Queen Charlotte Is. and between Kodiak Is. and Ocean Cape; Skud (1977) has, however, shown that the separation between the two stocks may no longer be tenable. The size of the filled circles indicates numbers of eggs found.

1930); in Area 2 in Hecate Strait there was a mean dispersion of 35 km, whereas in Area 3 off Alaska (more than 1600 km away) the mean dispersion of mature fish was 400 km, another stock difference. Another stock difference was established by egg survey (Thompson and van Cleve 1936); there were two groups of eggs, one off Queen Charlotte Is. and the other between Cap Spencer and Kodiak Is. (Figure 70).

The state of the stocks was described by Thompson and Bell (1934). Since 1907 stock density (as catch per six-line skate) had declined by five times in Area 2 and the number of skates rose by a factor of 3: 'control must be the regulation of the intensity of the fishing, whether the intensity is stated in terms of gear or proportion of stock removed'. In Area 2 the annual fishing mortality rate was probably about 40% year^{-1} and, in Area 3, 10% year^{-1}. But Thompson was most impressed by the inverse dependence of catch per unit on effort, as had been Garstang.

Thompson (1937) described his method. With constant recruitment, survivors were calculated at different annual rates of fishing and of natural mortality. Given the weights at age, the total weight of fish dying was calculated each year and the fraction due to fishing estimated – an early form of the catch equation (see Chapter 11). With an estimate of catchability, q $(= F/f$, where F is the instantaneous rate of fishing mortality and f is fishing effort, or time spent fishing), catch and stock density could be predicted from the quantity of fishing effort. Thompson (1936) wrote: 'had the growth exceeded the loss by death, the fishermen would have actually gained poundage by a less intense fishery because it would allow greater growth. And, on the other hand, had growth been less than death, the fishermen would have lost poundage'. Hence, regulation corrected a wasteful economic process, and also Petersen's view of growth overfishing (see above). Recruitment overfishing, 'the failure to leave enough adults to spawn' was not tackled, although Thompson recognized the problem. The model used resembles those in modern usage, but the objective was merely a stable yield.

Russell

Russell's major achievement after his visit to Raymond Pearl (a population expert who developed the logistic equation) was to state the axioms of fisheries science and to describe the simple dynamics (Russell 1931). The stock must be self-contained and fished systematically over a broad area. Russell distinguished between catchable stock and that which is not caught. He defined the catch rate as the number caught per unit fishing time by a standard vessel (hence changes in fishing power should be measured).

His famous equation states:

$$S_2 = S_1 + (A + G) - (C + M)$$

where

S_1 is the weight of stock at the beginning of the year and S_2 that at the end;

A is the sum in weight of the number reaching the minimum size of the catchable stock during the year, i.e. annual recruitment;

G is the annual increment in weight by growth;

C is the annual decrement in weight of catch;

M is the annual decrement in weight due to natural mortality.

Russell (1931) described S_1 and S_2 as the capital, $(A + G)$ as income and $(C + M)$ as expenditure. He investigated the variation in rates and how they might interact. He wrote that it was desirable to 'keep S at such a level, or to bring S to such a level, that the maximum value of commercially utilizable fish can be drawn from it annually without causing a progressive diminution of S'. But recruitment fluctuates and 'the ideal of a stabilized fishery yielding a constant maximum value is impracticable. It might . . . be practical politics to attempt to adjust the amount of fishing each year to the variations in the stocks of particular fish in particular regions . . . If such variations in abundance could be foretold a year or so in advance, this adjustment could be made more rapidly and with more certainty of success.' Written in 1931, this statement pre-dates the establishment of total allowable catches (TACs) in the Atlantic, by some forty years.

Thus the objective of a maximum sustainable yield was stated and the cure for growth overfishing was proposed in a preliminary way. Recruitment overfishing was considered to be unlikely as it was 'never actually met with so far as fish are concerned'. The approach differed from Thompson's in that there had been a long history of the capture of undersized fish, particularly plaice, in the North Sea. It is likely that Russell knew very well that the fleet was a more powerful sampling instrument than was a research vessel.

Graham

Graham (1935) started with two propositions: first, that with reduced death rate the average age of the stock should increase and, secondly, that there is a 'profitable age at which to harvest'. He wrote that yield was expressed as a flat-topped curve (as function of stock) and so 'it is most practical to say that it will pay to reduce fishing so long as yield is not thereby reduced'. He also noted that the maximum was 'not exactly the most profitable, . . . some further economy can still be made by reducing fishing, depending on the ratio of overhead costs to running costs per ton of fish'. The basic tenet of the paper was that the age of first capture be raised from 2.5 years to 3.5 years, i.e. by the introduction of mesh regulation; the British had introduced mesh regulation for the British fleet in 1933.

The first step in the argument was to note that the total yield from the North Sea in 1909–13, when more was discarded, was the same as in 1928–32. Graham concluded that, because of the discards, the yield would be the same if the age at first capture had been increased. He then constructed a small model, setting the rate of natural increase equivalent to the rate of fishing and showed that if the age at first capture was raised, a compensatory change in natural mortality was unlikely, 'so yield would be no less, were the fishing effort reduced so as to allow the fish to become one year older'.

Graham's (1935) main model states that the rate of natural increase varies as the difference between the weight of stock (P) and its maximum (P_m), which is of course the logistic equation as developed by Raymond Pearl (see above). He noted that the increment of stock density between 1913 and 1919 amounted to a factor of 2.08; this, of course, was the demonstration of the effect of fishing with the relaxation of fishing effort during the First World War. Graham (1935) concluded that 'the upper limit to stock is not less than twice the prewar weight'. He then estimated that the relaxation of effort corresponded to a respite of 2⅔ years. From tagging experiments the pre-war fishing rate was taken to be 30–50% year^{-1} and 50% year^{-1} in the 1920s and early 1930s. Figure 71 shows the model in graphical form; $P_m = 220$, $P_{1913} = 100$ and the annual fishing mortality, 40% year^{-1}. From age zero in the middle of the figure, the stock changes as differences from P_m were incremented or decremented in tenths of a year. The curve constructed in this way is one of stock increase showing its natural growth in time. The curve of yield was constructed from the differential coefficients, or slopes, of the stock curve.

The figure shows that the maximum stable catch was only a little above that in 1913 (A), so the drain from 1920 to 1933 could not be maintained (to B, the stock in 1933). The pre-war stock could have maintained a catch greater than that in 1933 by 14%. Hence, the maximum stable yield is

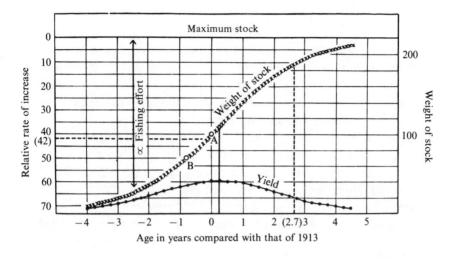

Figure 71. Graham's (1935) logistic model. The increase in stock during the First World War with a relaxation of fishing effort corresponding to a respite of 2.67 years. The relative rate of increase of stock is set equal to the inverse of the fishing effort. 'A' represents the stock in 1913 and 'B' that in 1933. The figure was constructed from 'A' (stock at 100) by taking increments of stock in time (tenths of a year) with respect fo a maximum stock of 220. The pre-1914 rate of fishing (from tagging experiments) was considered to be 40% per year and, in 1933, 50% per year. The yield curve was calculated from the differential coefficients (or slopes) of the curve in stock. Comparing the yield in 1913 with that in 1933, a gain of 14% could be obtained by reducing fishing effort by 10%.

15% greater than that in 1933 and could be taken by only 75% of the fishing effort in 1933.

Hjort *et al.* (1933) had shown that a maximum yield could be taken from a natural population. Graham's considerable achievement was to adapt the logistic equation to the age distribution of fishes and to the data collected on stock density and fishing effort. A maximum yield was established, as was the scientific justification for the increase in the age at first capture, or mesh size in the cod ends of the trawls. A solution of the problem of growth overfishing became conceivable although 'most fishermen and some scientists would also add "and a better chance to breed".' No progress was made internationally before the outbreak of the Second World War, but Graham's paper laid the foundations of the Overfishing Convention of 1946.

Conclusion

The origins of fisheries science lay in the marine biology of the nineteenth century; from the expeditions arose the intense study of systematics and the science of oceanography. It was the assurance of the sea-going biologists that led governments to seek solutions internationally, based on science, to the problems of the fishermen.

Petersen and Garstang did their major work in the last decade of the nineteenth century. Petersen faced the problem of stock identity and made the first, admittedly rough, attempt to age the stock from length distributions. He tagged fish and believed that the natural mortality of the Kattegat plaice was low. He distinguished between growth overfishing and recruitment overfishing. Garstang used the inverse relationship of stock density on fishing effort to show that North Sea trawled stocks had declined significantly in the last decade of the nineteenth century.

The International Council for the Exploration of the Sea was founded at the time when Petersen and Garstang were publishing their first results. But in the first two decades of its existence it failed. The reason was two-fold, that the statistics of small samples had not then been invented and that there was no theoretical basis to work from. The Council's statistician was unhelpful but the collection of statistics were started properly under the influence of d'Arcy Thompson.

In the 1920s, W. F. Thompson established the regulation of the stock of Pacific halibut. His science resembled Petersen's, with its insistence on stock identity, and was like Garstang's in its reliance on the inverse dependence of catch per unit of effort, or stock density, on fishing effort. The model he developed resembled a rough yield per recruit with no maximization. But his best achievement was an international regulation which worked, despite the controversy on natural changes and the poor economic outcome.

Russell was aware of the need for a maximum yield and of the use of mesh regulation to approach it. He stated the axioms of fisheries science in the clearest way and formulated the dynamics in a simple manner. Graham pursued the aim of mesh regulation and applied the logistic equation to establish the maximum sustainable yield. In forty years fisheries science had acquired a theory.

10

Institutions before
the Second World War

Fish stocks are often shared between nations. The need for international cooperation and perhaps common regulation was recognized soon after the evidence of stock depletion. Institutions appeared in Europe and in the north-east Pacific in the first two decades of the present century. The International Council for the Exploration of the Sea (ICES) was established in Copenhagen in 1902. In the north-east Pacific, three species commissions were set up, the North Pacific Fur Seal Commission in 1911, the International Fisheries Commission in 1923 (later, in 1953, the International Pacific Halibut Commission) and in 1937 the International Pacific Salmon Commission. The three species commissions were of North American origin (i.e. Canada and the United States), but the Russians and Japanese took part in the Fur Seal Commission. Each commission was established by treaty and any regulation would be first agreed and then enforced on the basis of scientific advice supplied by scientists employed by the commission. The International Council was (and is) primarily a forum for the provision of scientific advice, although it could at that time proceed to negotiate between nations through the Danish Foreign Office. Such were the institutions, and in the period before the Second World War they took their first hesitant steps towards the management of fisheries on an international scale.

The International Council for the Exploration of the Sea

The small plaice problem

As noted in Chapter 9, one of the first actions taken by the International Council was a study of the overfishing problem. The latter had been described by Petersen (1894, 1900) and by Garstang (1900). Petersen had distinguished between recruitment and growth overfishing; the former occurs when recruitment is reduced as a consequence of declining spawning stock biomass. Growth overfishing takes place when young fish are

killed before they had been able to put on their best weight for the total
yield. Petersen thought that fishermen would solve the problem of growth
overfishing by themselves if it was explained to them. Garstang showed
that the catch per unit of effort (where effort is time spent fishing) of the
British North Sea trawl fleet declined by between a third and a half as the
number of steam trawlers increased. The catch per unit of effort, or stock
density, is of course also the average catch of the vessel, catch per day's
absence or catch per 100 h fishing. It is the index by which the skipper
marks his success in fishing and much of the pressure for the establishment of ICES came from fishermen who knew that stocks were declining.

Also as noted in the last chapter, the early scientists were unable to use
the records from the research trawlers. This was long before Fisher had
introduced the statistics of small samples (Fisher 1925) and the high
variability of the material dismayed everyone. On the basis of the work of
Garstang (1909), Heincke (1913) showed that bigger plaice lived in
deeper water and that the small plaice lived close to the European coast.
In the shallow waters the trawlers in summertime were taking between
$500 \, h^{-1}$ and $1000 \, h^{-1}$ and they were often much less than 20 cm in length.
From early tagging experiments, Heincke concluded that the plaice
suffered an annual mortality of 20–50% and he concluded that the 'decrease of the large plaice and increase of small in the landings and the
corresponding reduction in the average size of the plaice are a direct
result of the more intensive fishing'. He proposed a minimum landed size
'as a first attempt, an experiment'. The outbreak of the First World War
prevented any further development, and Heincke did not return to the
Council after that war. In 1923 it was proposed that the areas where the
small plaice were abundant should be closed to fishing. The Belgian,
Danish, French, Swedish and Dutch governments desired an international convention but the British and Norwegian governments did not
reply to this proposal.

The Baltic plaice

By 1928 the stock densities of plaice in the Baltic had fallen by a factor of
between 5 and 10 since the end of the First World War and the growth rate
had increased in a density-dependent manner (Blegvad 1928; Andersson
and Molander 1928). Further, the catches themselves had decreased. The
International Council proposed a closed area during the spawning season
and a minimum landing size of 24 cm. Through the Danish Foreign
Office, the Council proposed the Convention for the protection of plaice
and flounder in the Baltic, the Baltic Convention. This proposal was put

to the governments of Danzig, Denmark, Germany, Poland and Sweden; in 1933 it was signed. It was the first international convention for the management of fish stocks in the Atlantic. The signatory governments were convinced by the scientific evidence presented to them by the Council on the decline of stock density and the decrease of total catch.

Science in the Council in the 1930s

Russell (1931) had formulated the 'overfishing' problem as the sum of the annual gains in weight by recruitment and growth less annual losses by death due to fishing and to natural causes (see Chapter 9). In 1932, there was a meeting in the International Council to discuss the effect of the capture of undersized fish upon the stock. With a model calculation, Bückmann (1932) suggested that with increased numbers (with relaxation of fishing) growth might be so depressed as to yield no increase in the weight of catch. At the same meeting, Russell (1932) noted that only the extra increment of the growth of survivors should be set against the loss in weight caused by thinning the stock; but Bückmann's point might have been valid for the subsequently recruiting year classes. Half a century later this reads strangely, because density-dependent growth in adult fishes is a little difficult to establish; it exists (see, for example, Houghton and Flatman (1981) on the North Sea cod and Southward (1967) and Deriso (1985) on the Pacific halibut), but there is a sense in which we believe today that adult fish are often well fed. The point is of historical importance because Russell's result allowed him to proceed.

Davis (1936) had shown that larger mesh sizes released the smaller fish to put on more growth before recapture (Figure 72). Russell noted in 1934 that minimum landing sizes and a mesh regulation had been introduced in Great Britain. In March 1937, a conference was held in London on the protection of undersized fish: a convention was signed by ten countries but a ratification was prevented by the Second World War.

As described earlier, Graham (1935, 1939) formulated a model of the exploitation of a stock by fishing. Thompson and Bell (1934) had shown an inverse dependence of stock density upon fishing effort (the time spent fishing) for the Pacific halibut stock, as indeed Garstang (1900) had shown for the British trawling fleet. Figure 73 shows how catches and stock densities of haddock and plaice changed during the First World War (Russell 1939). Stock densities of both species rose by a factor of 2–3 after the relaxation of fishing effort during that war. Catches in weight rose very quickly after the war, but peaked in two years, after which they declined. Subsequently, stock densities declined by a factor of 3 and total

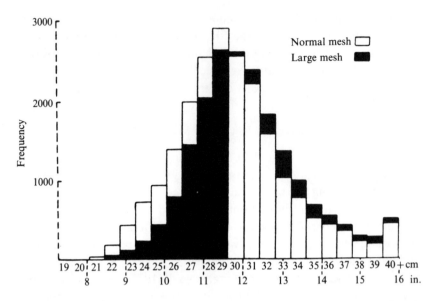

Figure 72. The increment of average size of fish caught with an increase of mesh size in the trawl (Davis 1934).

catches were reduced. It was this evidence that convinced people that there was a middle range of fishing at which the best capture might take place.

Following Heincke (1913), Graham (1938) had shown how to estimate fishing mortality from a tagging experiment. Plaice were tagged with vulcanite numbered discs secured with silver wire with Petersen's method (1896) and it was assumed that the tagged population was a small but representative sample of the real one. Then the proportion of tags recaptured in a time interval estimates the proportion of fish caught in the stock. In 1939, a meeting on overfishing was held by the International Council, at which Russell's evidence of the effect of the relaxed fishing effort during the First World War was presented (Figure 73(a)). Graham (1939) gave a paper on the theory of fishing, which is set out in his book *The fish gate* (1943). The Second World War prevented any further movement.

In 1946 the Overfishing Convention was held in London and it was signed by most European countries. It proposed minimum landing sizes for demersal fishes and minimum mesh sizes for trawls, except those used for herring and for shrimp, but there was no agreement to limit entry to a fishery. As the Convention became ratified, the Permanent Commission was established in 1954. It was the derivative and active organ of

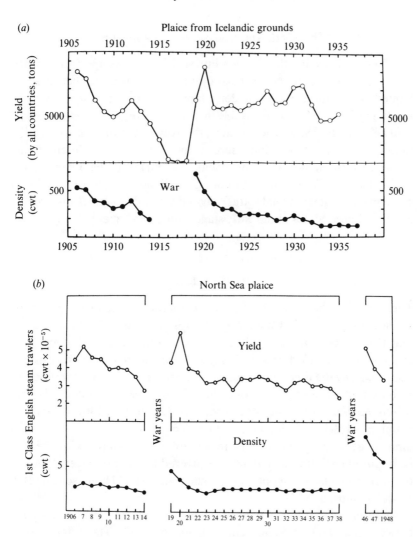

Figure 73. The relaxation of fishing effort during the two world wars. The change in catches of North Sea plaice and stock densities in cwt per day's absence: (*a*) before and after the First World War (Russell 1939); (*b*) after two world wars (Wimpenny 1953).

the Overfishing Convention. As will be shown below, the greatest yield is taken by allowing the little fish to escape from the trawl and put on weight before they are caught again. The political solution to the problem of growth overfishing was to agree minimum mesh sizes between the nations. This was the major achievement of Russell and Graham.

The North American species commissions

It will be recalled that the Pribilov Fur Seal Commission was established in 1911, following the North Pacific Fur Seal Convention between the United States, the United Kingdom, Japan and Russia. It was successful at a time when the only comparable international organization to manage wild animals, ICES, was not. Hence, the Americans probably wished to follow the Seal Commission's success by supporting the International Fisheries Commission (for the Pacific halibut) and the International Pacific Salmon Commission. In both Commissions, scientists employed by them (with international status) advised Commissioners (from Canada and the United States in the early stages) on management.

Pacific halibut

The history of the Pacific halibut and Thompson's role in advancing the science is described in Chapter 9. Thompson (1916, 1917a) pointed out that the stocks were exhaustible; he showed that the stock density, as catch per unit of effort, was declining and concluded that death by fishing was caused by the time spent fishing, or the fishing effort. He proposed an extension of a closed season which had already been suggested, a ban on fishing halibut on Hecate Strait (between Queen Charlotte Is. and the mainland of British Columbia), but much more important he proposed a limit to the number of boats and men employed. Such restraints were to be imposed until the stock density recovered to an optimal level, but he never revealed how that optimum should be defined (Thompson and Bell 1934; Thompson 1952).

An International Joint Commission met in 1918 and a treaty was signed between the United States and Canada in 1923. The International Fisheries Commission was established in 1924 and a midwinter closed season was enforced in November of that year. Thompson became the first director of the IFC. He arranged tagging experiments (Thompson and Herrington 1930) and egg surveys (Thompson and van Cleve 1936), as described in the previous chapter.

Stock density continued to decline between 1924 and 1930 because the midwinter closed season was ineffective. A new treaty was negotiated in 1930 and in that year two nursery areas were closed, a minimum landed size of 26 inches was enforced and the number of boats restricted by licences. As the years went by, the closed season was extended, but the important point was the limitation of entry. There has been long dis-

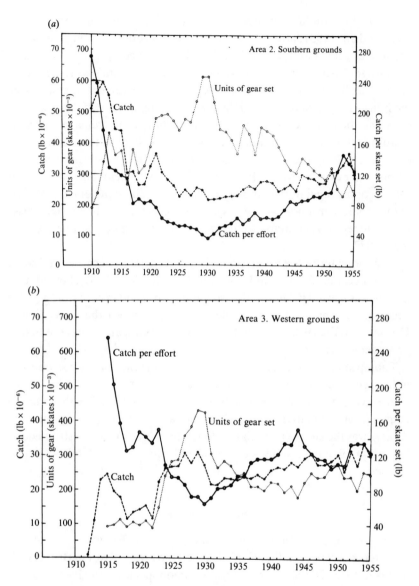

Figure 74. The recovery of the Pacific halibut; (*a*) southern area; (*b*) western area (INPFC 1962) (1 lb ≃ 0.4536 kg).

cussion on the need for limited entry in the half century since it was first introduced. It was Thompson who introduced it and it was successful.

Figure 74 shows the effect of regulation on the Pacific halibut fishery on the southern grounds (area 2, (*a*)) and on the western grounds (area 3, (*b*)). On the southern grounds the fishing effort as units of gear set increased until 1931, after which it declined as an effect of licensing. Stock density, as catch per unit of effort, increased immediately after 1930 and catches gradually increased subsequently. Such a result was too good to be true, as Burkenroad (1948) pointed out: the increment of stock after the relaxation of fishing effort must await the appearance of the next year class into the fishable stock, which may take a year or even a few years if recruitment is spread in years. However, stock density continued to increase and it is probable that a good year class appeared when the regulation was put into effect. If so, Thompson was lucky and the overall success of his regulation is shown in the figure. The same effect is shown in the western grounds (area 3), which suggests that the lucky year class, if it occurred, was common to the two stocks (which may not be distinct, Skud 1977).

In the long term, the regulation was less successful, because the fishing season became shorter and shorter; Crutchfield (1965) showed that money was wasted in laid-up fleets, maintenance and in cold storage. The licensing system precluded trawlers and encouraged the dying liners. However, in the 1930s the regulation was successful and it was the first to be so demonstrated.

Thompson (1952) said that he had been lucky: the fishery was uniform in fish, gear and fishermen (of Scandinavian origin whether in Seattle or Vancouver) and there were only two nations involved with common language and common law. But an equally important component of success was the simple formulation that if fishing effort was restrained, by limitation of entry, stock density would recover and later so would catches.

Pacific salmon

During the construction of the Trans-Canadian Railway, a landslide occurred at a canyon called Hell's Gate on the Fraser River in British Columbia in 1913. Of the stocks of the six salmon species in the Pacific northwest, that of the sockeye in the Fraser River is probably the most important. The decline of the Fraser sockeye dates from 1913. In 1937, the International Pacific Salmon Commission (IPSC) was established, and Canada and the United States provided Commissioners. The first director was W. F. Thompson and no action was to be taken until two of the four year cycles of spawning had elapsed.

Thompson showed that the spawning groups below Hell's Gate had not declined at all. Results from tagging experiments showed a clear relation-

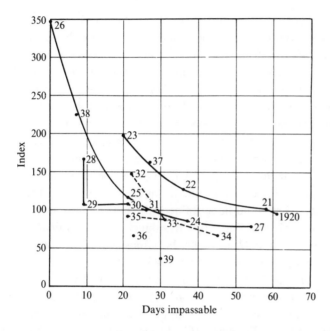

Figure 75. The index of return of the Fraser River sockeye salmon above and below Hell's Gate (Thompson 1945).

ship between tag recovery above Hell's Gate and the water level there. Thompson (1945) devised an 'index of return', the ratio of catch of spawners in a given year to that of their parents four years before; the index was weighted by differences in fishing effort. It is, of course, based on the discovery that the sockeye return to spawn and then die on their native redds (Foerster 1936; Pritchard 1939). Figure 75 shows the inverse relationships between Thompson's index of return and the number of days in the year in which Hell's Gate was impassable. So the decline in the Fraser River stock of sockeye was due to the landslide. In 1945 and 1946, a new fish ladder was built at Hell's Gate and by 1957, the weakest of the four annual cycles in the stock had increased by an order of magnitude. For a considerable number of years the fishery was closed for a period of 36–72 h per week until catch in each year did not exceed the index of return as four years before, and gradually the stock recovered.

Conclusion

At first sight there is a sharp contrast between the slow progress in ICES from its inception in 1902 to the Overfishing Convention in 1946 and the rapid achievement of results in two North American commissions. In a sense ICES was founded too early in the century, for the science was not published until the 1930s. Although Petersen and Garstang had sketched the nature of fisheries science at the turn of the century, it was the work of Russell, Thompson and Bell, and Graham which put it in quantitative form. In this light there was little difference between ICES and the North American commissions in coming to the right conclusions.

If the nations are to act in concert, the science must be expressed simply. Russell's simple equation described the dynamics, and Thompson expressed a principle of control in the inverse dependence of stock density upon fishing effort. Graham was the first to see that there was a greatest yield that could be sustained, the maximum sustained yield of Ricker (1946); his 1935 paper on this subject is a little difficult to read but in the meetings in 1937 and 1939 he almost certainly used the format presented in *The fish gate* (Graham 1943).

The outcome of the Overfishing Convention in 1946 was the establishment of the Permanent Commission in 1953 (later the North East Atlantic Fisheries Commission, NEAFC) and the International Commission for North West Atlantic Fisheries (ICNAF) in 1949. As will be shown below, these two commissions established a broad base for fisheries regulation throughout the North Atlantic. In contrast, the two North American species commissions did not extend their aegis and became, at least during the 1950s and 1960s, somewhat restricted. The reason for this was that the science of Thompson was retained without modification, in a somewhat too respectful manner.

11

Fisheries research 1946–65

The period of twenty years after the Second World War starts with the Overfishing Convention of 1946 and ends at the height of the expansion of fleets after the second industrialization (see Chapter 12). The date is arbitrary but during the middle 1960s the limits of mesh regulation on its own were becoming clear and the appearance off the shores of developing countries of fleets foreign to them was starting the movement towards the 200 mile limit. During this period the science was formed by Ricker, Beverton and Holt, and Schaefer. At this point the work of Baranov (1918) should be noted; in many ways he anticipated the science of the 1950s in that he devised an operational model of the North Sea plaice population in yield as a function of fishing mortality. His paper was translated by E. S. Russell in 1938, but had no influence. His model overestimated the importance of old fish at low fishing mortalities because length was said to increase proportionally with time, which leads to a cubic relationship between weight and age and overestimates yield.

Baranov had formulated the catch equation in numbers, as did Ricker (1948) and Beverton and Holt (1957). The number of deaths in a year class is given by the difference in numbers in the stock between the start and end of the year. The number of deaths during the year divided by the total mortality, Z, gives the average stock $\bar{N}$. Then the number of deaths due to fishing, or the catch in numbers, C, is given by:

$$C = F\bar{N}$$

where F is the instantaneous coefficient of fishing mortality. This expression can be applied to all catches by age in a year class or to all catches by age in an annual age distribution.

One of the basic axioms of fisheries science is that the instantaneous rate of fishing mortality is proportional to the fishing effort, f, the time spent fishing, i.e.:

$$F = qf$$

where q is the catchability coefficient. So,

$$C = qf\bar{N} \text{ and } C/f = q\bar{N}.$$

The catch per unit effort, the catch per unit of time spent fishing (the average catch of a fishing vessel each day – the skipper's index of success) is thus a proper index of the average stock in the sea. This is a most important statement, realized (but not always stated) by the prominent scientists working before the Second World War. However, as will be shown below there are biases in the estimation of catchability.

Fishing mortality was estimated by tagging experiments. Although Thompson and Herrington (1930) and Graham (1938) had formulated methods of estimation, Ricker (1948) introduced methods that are now standard. He assumed that the two forms of mortality, fishing and natural, were the same in the tagged as in the natural populations and that the tags were mixed uniformly throughout the natural population. Ricker described two forms of mortality due to tagging: type I soon after tagging (due to handling, shock etc.) and type II which persisted throughout the period of tagging (loss of tags etc.). At this time, Ricker worked in the lakes and streams of Indiana and the methods developed were suitable for seasonal tagging experiments in the quickly mixing waters of lakes. He also provided statistical methods for this work. All were based on the reasonable assumption that the tagged population was a representative sample of the wild one.

Ricker (1946) developed a method of estimating the yield in weight as a function of fishing mortality, on the basis of loss of numbers by mortality and an exponential increase in the weight of individuals. As in Baranov's method, this leads to an overestimate of older fish when mortality is low. Fish growth may be described exponentially for short periods, but for longer periods approaching that of the life cycle the specific growth rate decreases with age, as will be shown below. Although Ricker's method would not yield maxima of yield as functions of fishing mortality, he introduced the term *maximum sustained yield*, later changed a little to *maximum sustainable yield* (Chapman, Myhre and Southward 1962), a term which found its way into certain treaties. The term expresses a most important principle: that there is a greatest catch that can be maintained for long periods. There are two meanings to the phrase: (*a*) the maximum of the curve of yield in weight versus fishing mortality, which can be defined precisely; (*b*) a more general notion that there is a greatest catch that can be safely taken for a long time. Today we know that the technical maximum is undesirable and that we have to work at lower fishing mortalities for security and for the best economic yield. In the second and more general sense, this is what the fishermen would mean and what they desire.

The plaice population in the southern North Sea had been studied during the 1930s by Thursby-Pelham (1939); she used the annual rings on the otoliths (earstones used by fish to balance in the water) successfully to make age distributions in stock density by quarter years for a period of about ten years. Hence the changes in the age structure of the stock were readily portrayed. It had been possible to age fish during the first decade of the century (Dahl 1907), and Lea (1929) had described the structure of the Norwegian spring spawning herring population, particularly the passage of the great 1904 year class through the age distributions. But the importance of Thursby-Pelham's work lies in the use made of it by Beverton and Holt (1957).

They made two advances: (1) They used the age distributions to estimate the total mortality suffered from year to year by a year class from the logarithmic ratio of successive yearly estimates (weighted for abundance); their result is shown in Figure 76(*a*). (2) They applied the von Bertalannfy (1938) growth equation to describe the growth of plaice in length or weight as function of age; their result is shown in Figure 76(*b*). In length,

$$l_t = L_\infty \left(1 - \exp\left(- Kt\right)\right),$$

and in weight,

$$W_t = W_\infty \left(1 - \exp\left(- Kt\right)\right)^3,$$

where l_t and W_t are length and weight at time (age) t; L_∞ and W_∞ are asymptotic lengths and weights at infinite age; K is the rate at which the asymptote is approached in time t. Although other equations have been used to describe the growth of fishes for particular purposes, the von Bertalannfy equation remains that in most general use. Its advantage is that the reduction in specific growth rate with age in juvenile and adult age groups is well described but larval growth tends to be overestimated; an additional advantage lies in the ease with which it can be integrated into the yield equation (see below).

The main achievement of Beverton and Holt (1957) was the development of the expression for yield (in weight) per recruit. The magnitude of recruitment to any fish stock from year to year is highly variable, from a factor of 3 to perhaps two orders of magnitude. That variability is evaded in the expression of yield per recruit. The initial equation states that the rate of change of yield with time depends upon the product of biomass (numbers × weight) and fishing mortality. Because the rates of growth and of mortality in time differ, they solved a partial differential equation

(a)

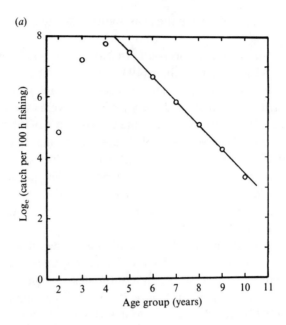

(b)

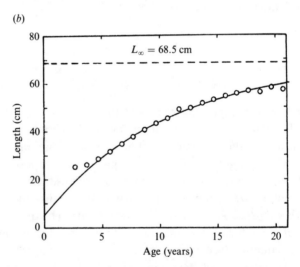

Figure 76. The use of age distributions of plaice between 1929 and 1938 to obtain: (a) the average rate of total mortality; (b) the average rate of growth (Beverton and Holt 1957).

in two stages and the integral throughout the fishable life span was:

$$Y_W = FR' \, W_\infty \, \sum_{n=0}^{n=3} \frac{\Omega_n \exp(-nK)(t_\varrho' - t_0)}{F + M + nK} \, [1 - \exp(-(Z + nK)\lambda)]$$

where

Y_W is yield in weight in g;

R' is the number of recruits to the fishable stock at age t_ϱ;

ϱ' is the age of recruitment to the fishery;

Ω_n is the expanded cubic of the equation for growth in weight;

$n = 0, 1, 2$ or 3, where $\Omega_n = +1, -3, +3$ or -1, respectively;

t_0 is the age at zero weight estimated from the growth equation (it is often negative because only the larger animals of the youngest age groups are sampled);

λ is the number of age groups in the exploited stock.

The yield-per-recruit equation is expressed in terms of Y_W/R'. When plotted against fishing mortality, Y_W/R' reaches a maximum or an asymptote, in F.

Having devised this relationship, all that remained to do was to estimate F and M, Z being already known. The natural mortality rate was estimated from the ratio of the trans-wartime year classes at the start and end of the Second World War ($\simeq 0.1$); it remains one of the best measures of natural mortality. Beverton and Holt (1957) used two methods of estimating fishing mortality. (1) From tagging: the subpopulation of tagged fish was examined from year to year (as opposed to Ricker's within-year methods, which may be more suitable to lake populations); from the ratio of tags returned in a given year to those recovered in the previous year, it was possible to separate the fishing mortality from the other loss coefficient (i.e. those of type I and type II referred to above) and hence to estimate the fishing mortality. (2) They plotted the ratio of successive stock densities in year classes against fishing intensity (fishing effort per unit area) in the first year of the pair; the regression so derived yielded an estimate of q from the slope and an estimate of natural mortality from the intercept.

Having obtained estimates of fishing mortality from tagging experiments executed before the Second World War they were able to construct the yield per recruit curve as function of fishing mortality (Figure 77). The pre-Second World War estimate of $F = 0.73$ and the maximum to the yield per recruit curve is shown at a value of about $F = 0.22$. In other words, the yield per recruit could be increased by nearly one-third if two-thirds of the fishermen were put out of work.

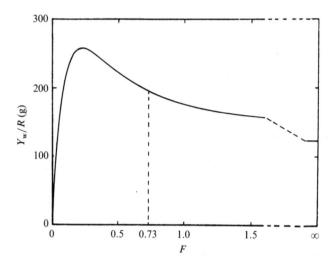

Figure 77. The yield in weight (y_w) per recruit (R) of the North Sea plaice as a function of fishing mortality (F) (Beverton and Holt 1957).

Beverton and Holt avoided this impossible solution by rearranging the relationship in terms of age at first capture (which is effectively mesh size in the cod ends of the trawls). Figure 78 shows an isopleth of age at first capture against fishing mortality. It is obvious that as the age of first capture is raised so the yield per recruit is raised for a given fishing mortality. Thus, the solution to the problem of growth overfishing was to increase the mesh size to obtain greater yields at constant fishing mortality. Hence, the increment of yield could be obtained without putting fishermen out of work. The solution was a triumph because it laid the foundations of international regulation in the demersal fisheries in the North Atlantic during the 1950s and early 1960s. However, the figure also shows that, at constant age at first capture (or mesh size), if fishing mortality is increase there may be an increase in yield with further increments of fishing mortality.

The achievement of Beverton and Holt was two-fold: (*a*) the use of the age distributions to describe properly the loss in numbers with age and the gain in weight with age; (*b*) the development of the yield (in weight) per recruit as function of fishing mortality; this procedure remains at the heart of much of present day population analysis in fisheries. The reason for this is that, once the curve is constructed, decisions can be made on action and frequently the curve can be derived fairly quickly.

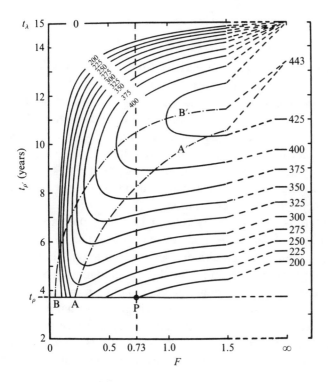

Figure 78. The yield isopleth diagram in which yield per recruit is given as function of age at first capture ($t_{\varrho'}$, or mean size) and fishing mortality (F) (Beverton and Holt 1957).

Schaefer (1954) returned to the logistic equation used earlier by Graham (1935). The difference between maximum stock, P_m, and present stock, P (i.e. $P_m - P$), is inversely related to stock and is a function of the stock's natural rate of increase. The catch taken from the stock equal to the natural rate of increase is the equilibrium catch, C_e. Schaefer (1957) derived a difference equation which stated that the increment or decrement in stock, ΔP, varied as the equilibrium catch less the catch, C (i.e. $\Delta P = C_e - C$).

The catch per unit of effort, the stock density, is a proper index of stock, $\bar{U}$, averaged over a year (i.e. $\bar{U} = q\bar{P}$, where q is the catchability coefficient and $\bar{P}$ is the average stock in a year). Schaefer then expressed his difference equation in stock density as $\Delta\bar{P} = \Delta(\bar{U}/q)$. Then:

$$C_e = a\bar{U}(M-\bar{U}),$$

where M is the maximal catch per unit of effort; $\bar{U}$ is the average catch per unit of effort in a given year; a is a constant. So

$$C_e/\bar{U} = a(M - \bar{U}).$$

From estimates of C, $\bar{U}$, $(C/\bar{U})$, ΔU and $\Delta U/\bar{C}$ for each year, Schaefer (1957) devised statistical methods to obtain estimates of a and M, and so he could draw a curve of equilibrium yield, C_e, on fishing effort. Figure 79 shows this curve for the Eastern Pacific stock of yellowfin tuna. The curve has confidence limits and a maximum sustainable yield (or 'maximum average equilibrium catch') was estimated. Although this particular estimate was replaced by a larger one in subsequent years, the method demands only the statistics of catch and catch per unit of effort; further, there is no need for age determination.

Because $(P_m - \bar{P})$ is inversely related to stock, the rate of natural increase must respond immediately to change in stock. In the stock this rate includes changes in recruitment, growth and natural mortality, and Schaefer did not try to distinguish the three components. Today we tend to believe that the lagged response in recruitment is more important than the immediate density-dependent changes in growth or natural mortality. Also, the increment in weight of a recruiting year class is relatively greater in an exploited stock than in an unexploited one. However, the rate of natural increase forms a parabola in stock with a maximum at half stock.

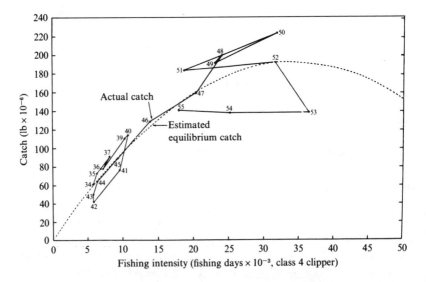

Figure 79. The yield of yellowfin tuna in the eastern tropical Pacific as function of fishing effort, using the Schaefer method; the maximum sustainable yield, with errors, is shown (Schaefer 1957). (1 lb = 0.4536 kg.)

This implies that, when stock is reduced below half, recruitment is reduced even if lagged. If the maximum yield is sustainable, the reduction in recruitment must occur at some point less than half stock. Hence, the Schaefer model if well applied should protect the stock from recruitment overfishing.

The dependence of catch in weight per unit of effort on effort is not linear as Thompson (1952) showed; for the yellowfin tuna, however, Schaefer (1957) demonstrated within the range of fishing effort observed that the linear fit was adequate. Further the parabolic curve may not represent the yield from an age-structured population (Schaefer and Beverton 1963); such a curve might well be skewed to the left.

Today there is a more serious potential defect. The catchability coefficient may vary inversely with abundance (Pope and Garrod 1975; Ulltang 1976; MacCall 1976). Then the true maximum sustainable yield is less than that estimated with Schaefer's methods. The effect is perhaps limited to shoaling fish vulnerable to searching fisheries (like the Norwegian herring or the Arcto-Norwegian cod). If that relationship is known or known to be absent, the Schaefer method remains useful where there are only observations of catch or of stock density. Perhaps Gulland's (1961a) simplification is useful: $Y = af - bf^2$, where Y is catch in weight, and a and b are constants from the linear dependence of catch on fishing ($Y = a - bf$). Gulland recommended that the annual data be averaged in groups of years, each of which represents half the effective age distribution.

The last stage in the development of fisheries science in the period 1946–65 was in the study of the dependence of recruitment upon parent stock, the stock–recruitment problem. Although Petersen believed that the reduction of recruitment by fishing down the stock was really the only form of overfishing, neither Thompson nor Graham believed that it would happen. Since the 1950s, many stocks have suffered from recruitment overfishing (as shown in Chapter 12). However, the first formulations by Ricker (1954, 1958) and by Beverton and Holt (1957) were written before such collapses took place.

Animal populations can be regulated by density-dependent changes in juvenile mortality or in adult fecundity. Both Ricker and Beverton and Holt rejected the possibility of density-dependent fecundity, although we now know that it can take place (for example, in perch in Windermere; Le Cren, 1958). In his 1954 paper, Ricker wrote: 'opportunities for compensatory effects are so much greater during the small, vulnerable stages of a fish's life that restriction of compensation to those stages seems likely to have wide applicability as a useful approximation'.

The Ricker equation depends on the assumption that the abundance of predators increases with the numbers of prey; that is, the predators aggregate on the prey. Then mortality has both density-dependent and density-independent components. The equation is sometimes written:

$$R = aP \, (\exp(-bP)),$$

where R is recruitment in numbers, P is stock in eggs, a is a coefficient of density-independent survival, and b is a coefficient of density-dependent mortality. An alternative assumption states that, if food is short, predatory mortality endures for longer, which is then density dependent. The curve of recruitment on parent stock is dome shaped and some examples (e.g. George's Bank haddock, Skeena salmon), given by Ricker (1954) appear to justify this conclusion (Figure 80(a)). But it implies that at high stock the density-dependent mortality is very high, which further implies that the initial effect of fishing must be to increase the magnitude of recruitment.

The Beverton and Holt equation starts from a different premise, that there is competition for food at a critical stage, for example when the yolk sac is exhausted (although they did not specify the stage). Such a mechanism is density dependent and it was assumed that lack of food causes mortality. Their equation is:

$$R = 1/(\alpha + \beta/E),$$

where R is recruitment in numbers, E is stock in eggs, and α and β are constants.

The curve of recruitment on parent stock is asymptotic and the observations on the North Sea plaice stock appear to justify this conclusion (Figure 80 (b)). There is then no implication that recruitment increases in the first stages of a fishery. The two formulations of the stock and recruitment problem have alternative biological explanations. No information has appeared since to make the choice between them (but see Elliott 1985).

Although many of the fundamental steps were taken in the 1930s, the firm foundations were not laid until the 1950s. Because the results were expressed in relatively simple terms, the international commissions were able to use them. As a consequence, during the 1950s and 1960s fisheries came under international control by voluntary agreement between nations, at least in the North Atlantic and in parts of the North Pacific. Today we may believe that the steps made in international cooperation were limited and that in fact the improvements to the fisheries were not very striking. One local result was that the plaice fishermen from

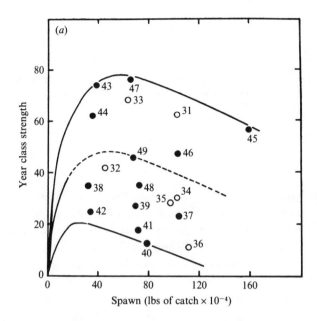

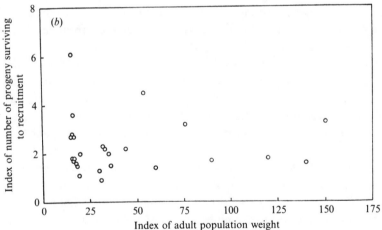

Figure 80. The dependence of recruitment on parent stock: (*a*) Skeena sockeye salmon (Ricker 1954); (*b*) North Sea plaice (Beverton 1962).

Lowestoft were told that the best form of conservation was to search for the larger fish, much bigger than those specified by minimum landing size and by the mesh size in the cod ends of the trawls. This they did for perhaps two decades – but a natural change had occurred which increased the German Bight stock of plaice and took Lowestoft fishermen almost to the coast of Norway.

12

Institutions between 1945 and 1965

The institutions in the North Atlantic during 1945–65 arose from the Overfishing Convention of 1946 and were established on the basis of the science of Russell and Graham. By the time that the more exact science of Beverton and Holt was published in 1957, the two Atlantic commissions, the Permanent Commission and the International Commission for North West Atlantic Fisheries, were in place. The first task of the two commissions was to regulate the demersal trawl fisheries on both sides of the Atlantic by mesh regulation. Growth overfishing was to be avoided by increasing the age at first capture, by enlarging the meshes in the cod ends of the trawls. Much of the progress made stemmed from the British experience. Figure 72 (p. 206) shows the effect of an enlarged mesh size on the length distribution of trawl-caught fish, i.e. a fair proportion of the smaller fish were let through the meshes of the cod end. Then it will be recalled that the 1934 Act in Britain increased the mesh sizes to 70 mm and that increment was linked to minimum landed sizes. The London Convention of 1937 on the protection of undersized fish was signed by ten countries but was not ratified before the outbreak of the Second World War. The Overfishing Convention of 1946 took up the thread again. The two commissions acted by voluntary agreement between nations and had no power to limit entry to the fishery, as did the Halibut Commission.

It is of some interest to examine the management objectives during this period.

(1) The International Fisheries Convention of 1946 (the Overfishing Convention) was 'convened for the purpose of examining the question of overfishing in the North Sea and other areas adjacent to the British Isles and if possible drawing up some form of regulatory agreement among the nations concerned' (Final Act and Convention of the International Overfishing Conference; Cmd 6791 (1946)).

(2) ICNAF was founded in Washington in 1949: 'a Convention for the investigation, protection and conservation of the fisheries of the North West Atlantic Ocean in order to make possible of a maximum

sustained catch from these fisheries . . . ' (International Convention for North-West Atlantic Fisheries; Cmd 8071 (1950)).

(3) The Permanent Commission (later NEAFC) was established in 1954 'to ensure the conservation of the fish stocks and the rational exploitation of the fisheries of the North-East Atlantic Ocean and waters, adjacent to them' (The North East Atlantic Fisheries Convention; Cmd 659 (1959)).

There is a history in the order of these three events. In 1946 the meaning of the word 'overfishing' must have been quite plain. In 1949, Ricker's 'maximum sustained yield' was enough to ensure conservation. In 1954 conservation was to be achieved by 'rational exploitation'. Then in 1958 the Law of the Sea Conference stated that 'the expression "conservation of the living resources of the high seas" means the aggregate of the measures rendering possible the optimum sustainable yield from those resources . . . ' In a decade, the word 'overfishing' had been replaced by a phrase which implied that something less than the maximum sustainable yield would secure good management. Today we accept a number of management objectives which may conflict. Earlier, the view was simpler and there is a sense in which that overall simple objective should be retained and within which conflicting and subordinate objectives could be resolved.

The Atlantic commissions

The North East Atlantic Fisheries Commission

The Permanent Commission in 1959 became the North East Atlantic Fisheries Commission (NEAFC; ratified in 1964). The change in name was associated with more extensive powers for the regulation of fisheries including power to regulate those for herring and shrimp.

The first task of the Commission was to extend the principle of mesh regulation to all the waters under its control. Today there is much discussion of multispecies stock assessment; part, but not all, of this problem lies in the fact that a trawl catches a number of species, three in the Barents Sea, six in the North Sea, eleven in the Irish and more than fifty in the Gulf of Thailand. An *ad-hoc* committee in 1955 (Anon. 1957) reported on the long-term effect of mesh changes in the North Sea:

the total weight of catch would increase for all species, if the minimum mesh size were increased. For sole and whiting, minimum mesh sizes of 80–100 mm were proposed, 100 mm for haddock and hake, and 140 mm for plaice and cod. So a common mesh regulation is really that for the smallest economically useful fish and larger fish will be exploited less efficiently.

Fishermen do not really like having to increase the mesh sizes in the cod ends of their trawls. There are two reasons for this: (*a*) they suffer an immediate loss when the new mesh size is introduced, which can only be mitigated by the appearance of a strong year class in the year of introduction; (*b*) many would prefer to see a large catch of small fish on deck (most of which might be discarded) to a smaller catch of larger and more valuable fish. This is in contradistinction to the experience of the Lowestoft fishermen, as noted above, who were persuaded to search for the larger fish.

A later committee (Anon. 1960) examined three problems. First, the fisheries for sole and whiting required particular arrangements and some small fisheries were allowed to use smaller meshes for a few years. A second effect arose because industrial fisheries for immature herring with small-meshed nets had started in the eastern North Sea during the 1950s and inevitably some of the protected species were caught. The catches were reduced to fish meal used for feeding animals. The catch of protected species was called a by-catch and the Committee recommended that such by-catches should not exceed 10% the total catch in weight. In later years the industrial fishery was to cause considerable problems, as will be described below. Thirdly and lastly, various problems of selectivity by twines of different synthetic materials were examined and selection factors estimated.

During the early 1960s the industrial fisheries were extended to small gadoid fish, Norway pout in particular. These animals lived on the same grounds as young haddock and young whiting, and conflict between the need to protect such species and the need to catch Norway pout for fish meal was inevitable. The 10% by-catch rule was retained against considerable efforts to increase it to higher proportions. Further, it was decided that there should be no nets of mesh sizes between 50 and 70 mm; intermediate mesh sizes had been used for catching *Nephrops*, the Dublin Bay prawn, in the Irish Sea. Although the rationale of this rule was sensible, it had the long-term effect of reducing the mesh size of the prawn trawls to 40 mm, which was unnecessarily low.

In January 1972, the more important mesh sizes in the north-east Atlantic were:

Area	Mesh (mm)
Bay of Biscay	70 mm
North Sea	80 mm
Faroe Is.	110 mm
Barents Sea, Iceland	130 mm

The short history of mesh regulation shows: (1) that its effect is limited to the smallest economically useful fish; (2) that in the North Sea the industrial fishery prevented the best exploitation of haddock, whiting and perhaps herring; (3) that in the Irish Sea and the Celtic Sea the exploitation of *Nephrops* was worsened rather than improved. The reason for this is that the mixture of species in a trawl catch encourages a least regulation rather than an optimal one. The justification for the multispecies approach lies in the hope that such conflicts can be resolved.

During this early period of the life of NEAFC, the East Anglian herring fishery failed; during the subsequent decade the Downs stock declined and the fishery for herring in the southern North Sea practically disappeared (Cushing 1968*a*). Hodgson (1957) explained the collapse of the East Anglian herring as a loss of recruiting herring in the industrial fishery east of the Dogger Bank. Recruitment had been reduced and Hodgson's explanation demanded a fishing rate of 60% year^{-1} in the industrial fishery. An international tagging experiment organized by the International Council for the Exploration of the Sea showed that the fishing rate was as low as 15–20% year^{-1} (Aasen *et al.* 1961).

It was suggested later that the Downs stock had suffered from recruitment overfishing (Cushing and Bridger 1966). Stocks of herring which spawned on the Dogger and in the northern North Sea probably suffered the same fate. A second tagging experiment was executed on the stocks of immature herring in the eastern North Sea in 1969 and 1970. The fishing rate was found to be high (higher than in the earlier experiment) as might be expected if the stocks had suffered from recruitment overfishing. An attempt made to conserve the herring of the southern North Sea failed to convince NEAFC in 1967, where it was rejected on scientific grounds. In later years fishing on herring in the North Sea was banned for a period and the Downs stock recovered. The case for the ban was that all the North Sea stocks had suffered from recruitment overfishing. In the late 1950s and early 1960s, it was thought that the problem was heavy exploitation of the adult stock and that the industrial fishery for immature herring was not entirely responsible; indeed we thought that it was less responsible. As recruitment became reduced by whatever cause, the role of the industrial fishery became crucial because its fishing rate was much increased. At the least it prevented the recovery of the herring stocks.

In the early stages of the NEAFC the intention was to solve the problem of growth overfishing by mesh regulation, in so far as it was possible. To a considerable degree this objective was achieved. However, the problems associated with the industrial fishery in the North Sea and with the *Nephrops* fishery in the Irish and Celtic Seas had already appeared; the subsequent developments will be described in Chapter 11. The collapse of the North Sea herring fisheries started during the period under discussion; there were two contributory factors, first that recruitment overfishing was at the time almost inconceivable and secondly that recruitment to the northern North Sea stock of herring was higher than had been recorded since 1920 (see Figure 46, p. 110). Subsequently a number of pelagic fish stocks have collapsed through recruitment overfishing.

The International Commission for North West Atlantic Fisheries

The ICNAF was established in 1949. In 1953 a 4.5 inch (115 mm) mesh was recommended for the haddock fishery on Georges Bank. A fleet of 'small mesh study boats' found that nets with larger meshes caught bigger fish and, in addition, the proportion of discards was reduced (Anon. 1953). Beverton and Hodder (1962) proposed that minimum mesh sizes be enforced in areas other than George's Bank. In later years they were indeed enforced, together with minimum landed sizes; by 1968, a complete system had been established.

There was no adequate time series of catch and effort statistics in the region and Gulland (1961*b*) devised a 'break-even' exploitation rate which had to be greater than the ratio of weight at first capture to mean weight. Consequently the long-term changes expected from an increase in mesh size could be predicted.

Scientific advice was provided by the Research and Statistics Committee; this means that a particular stock might be examined by groups of scientists which differed to some degree from year to year (in contrast the ICES Working Groups (see p. 288) tended to be composed of the same people from year to year, to the disadvantage of the science). During the 1960s the ICNAF meetings provided a scientific centre in the North Atlantic perhaps because of this difference in structure.

In 1965, ICNAF held an Environmental Symposium in Rome (Anon. 1965), where the broader issues of the biology of fishes in the exploited stocks were discussed. In the region between Cape Cod and the Grand Banks, environmental changes make themselves very obvious because of the conflict between the warm Gulf Stream and the cold Labrador current. In the north-east Atlantic analogous changes occur, but perhaps more slowly. One result of this symposium was to direct attention to the variability of recruitment and its possible causes.

Another study initiated by ICNAF was on the economics of fisheries. Economists had expressed the effects of exploitation in terms of cost and value and their formulations resembled the logistic equation (see

Chapter 11). But on the international scene, such efforts were baulked by the inability to transfer cost and value across the exchanges. However, the study led to some economic and biological work on purely national stocks, for example the Maine lobster. In a later period, methods were indeed derived to express the fact that the maximum economic yield is taken at levels of fishing less (or considerably less) than those needed for the maximum sustainable yield.

The Grand Banks and those off Nova Scotia had been exploited by European fishermen for centuries, as described in Chapter 4. During the 1950s and 1960s trawlers from the USSR and other East European countries started to work in larger numbers in the region. They worked on species not used by North American fishermen, such as hake, but also caught herring and in some years, haddock. This addition to the fishing effort in the region led to considerable problems in later years.

The Pacific commissions

The older commissions, IFC and IPSC (see pp. 208 and 210), remained as research institutions and for negotiations between Canada and the United States but their trans-pacific role was taken over by the International North Pacific Fisheries Commission (INPFC). A new commission was established in 1949, the Inter-American Tropical Tuna Commission (IATTC).

The International North Pacific Fisheries Commission

The INPFC was established in 1950. The abstention principle was its main instrument of conservation: a potential entrant must abstain from fishing if the maximum sustainable yield of the stock has been reached. The commission was established just before the peace terms were signed between Japan and the United States.

The Canadians agreed not to exploit the Pacific halibut in the Bering Sea when the abstention principle was applied in 1958. The Japanese abstained from further exploitation of the halibut, but in fact an agreement was not reached (INPFC 1962). Earlier they had agreed to abstain from catching herring off the North American coast until 1958. The South Alaskan stock of herring was removed from the *Annex* to the INPFC in 1960; the *Annex* lists the stocks from which Japanese fishermen abstained. In 1963, the herring stock south of the Strait of Juan de Fuca and that west of Queen Charlotte Is. were removed from the list. During this period the Japanese also abstained from exploiting the British Columbian herring.

Roughly, the British Columbian salmon live in the Alaska gyre, the six species of Alaska salmon live in the Bering Sea and the Kamchatka salmon live in the north-west Pacific gyre. The Japanese exploit the Kamchatka salmon in the open sea west of 175° W. They agreed, in 1950, not to fish for salmon east of that line and the Canadians agreed not to fish for salmon in the Bering Sea. Such agreements were the result of a considerable scientific effort. The degree of overlap between areas was investigated with a number of methods: experimental catches in different areas (Manzer *et al.* 1965), in the distribution of tag recoveries (Margolis *et al.* 1966), in the distribution of parasites (Margolis 1963) and by the discriminatory analysis of morphometric characters (Fukuhara *et al.* 1962; Margolis *et al.* 1966). It was the fairly low degree of intermingling that led to the agreements.

If abstention were to be agreed, the North American salmon stocks should be exploited at or near their maxima. Many curves relating recruitment to parent stock were presented, in the commission, for the Fraser, Skeena and Rivers sockeye, the Fraser pink and for the salmon stocks in Bristol Bay. Most showed that increased stock would yield increased recruitment so greater catches would be obtained with less fishing. Tanaka (1962) answered the case to some degree but the abstention principle was sustained.

The INPFC concerned itself with activities outside the scope of the older commissions. A minimum landed size was introduced for Bering Sea halibut and closed seasons were established for long-lining in particular areas. There were discussions on the king crab stock in the eastern Bering Sea but no agreement was reached within the commission; however, a bilateral agreement was signed between Japan and the United States.

The Inter-American Tropical Tuna Commission

The IATTC was established in 1949 to investigate the biology and population dynamics of the anchovy, the live bait for American tuna vessels. In 1947, the stocks of anchovy in the Gulf of Nicoya in Costa Rica collapsed. Costa Rica and the United States were the first members and they were joined later by Mexico, Panama, Ecuador and Columbia. Dr M. B. Schaefer, a pupil of W. F. Thompson (p. 196), was the first director. During the 1950s, the tuna boats stopped using live bait and the anchovy problem was replaced by that of the effect of fishing on the yellowfin tuna in the eastern tropical Pacific.

The unity of the yellowfin stock was established on the basis of morpho-metric measurements; differences were found between fish from Central America, Polynesia and Hawaii (Schaefer 1955). Within the area of the eastern tropics Pacific no such differences were found (Broadhead 1959). The spawning ground lies off the Revilla Gigedos Is. between May and September (Schaefer and Orange 1956), and the larvae and juveniles were found off the coasts of Central America (Klawe 1963). From an extensive tagging experiment, the most westerly recovery was taken 400 km west of the Galapagos Is. (Fink and Bayliff 1970). However, Royce (1964) demonstrated a cline of yellowfin along the equator and with morphometric measurements estimated an overlap of 50% in 2400 km. This suggests that the eastern tropical Pacific stock of yellowfin tuna is part of a larger group, within which there is interchange.

The major achievement of the commission during this period was the establishment of a catch quota in 1961 of 85000 tonnes. Schaefer's method was described in Chapter 11. In later years (1972) the catch quota had reached 120000 tonnes as the fishermen had searched further away from port. Perhaps the stock was in fact multiple, as Royce (1964) had suggested.

The commission scientists studied problems other than the stock of yellowfin. They described the mechanisms of upwelling in the Gulf of Tehuantepec. They established the dependence of skipjack catches on temperature (Broadhead and Barrett 1964). They took part in the oceanographic surveys at Scripps Institute of Oceanography and one of them discovered the equatorial undercurrent in the Pacific, the Cromwell current.

The Pacific commissions

The older commissions, for halibut and salmon, survived and their scien-tists worked in INPFC. Their original directors were outstanding fisheries biologists and they produced clear solutions which convinced the com-missioners quickly. But science often changes more quickly than scien-tists and the older commissions tended to preserve the science of their founding directors.

The INPFC conducts the scientific arguments openly in the commission and the quality of the *Bulletins* published is high. But the overall purview of the commission was restricted: only the North Americans and the Japanese were represented. Even within the small group, not all stocks are examined. During the period up to 1965, tuna in the North Pacific remained unregulated, as were the general demersal stocks across the Bering Sea. Indeed the fishery for Alaska pollack was to develop into the largest demersal one in the world for a single species.

There is a contrast between the Pacific species commissions and those in the Atlantic. The latter were established as the consequence of the political will expressed by the nations bordering the ocean; it was expressed in the Overfishing Convention of 1946 and in the establishment of ICNAF and NEAFC. The three species commissions in the Pacific were founded to protect American interests, as was the Pribilov Fur Seal Commission much earlier. Indeed the INPFC really originated in the American need to restrain the Japanese fishermen. This explains why in the first stages, at least, this Commission's purview was limited. It should, however, be remembered that a Northwest Pacific Commission resolves differences between Japan and the USSR.

The FAO commissions

During the short period under review, 1946–65, the Food and Agriculture Organization (FAO) of the United Nations established a chain of commissions throughout the world. Examples are the Indo-Pacific Fisheries Council and the General Fisheries Commission for the Mediterranean. Others were established in different regions, off West Africa, in the south-west and south-east Atlantic. Such commissions were advisory, which means that scientific advice is provided for the benefit of the participating nations. Initially, FAO provided technical and secretarial assistance. In general, however, the nations did not take powers to enforce any decisions, if agreed voluntarily. Consequently the commissions have become scientific forums, rather like the International Council for the Exploration of the Sea. In this role they have been successful. Indeed the Indo-Pacific Fisheries Council provided a scientific centre for a very broad area.

Conclusion

Before the Second World War, two fish stocks had been managed internationally, the Pacific halibut and the Baltic plaice. After the war, the stage was set for international management on a broader scale. The stimulus was the solution to the problem of growth overfishing, first in the hands of Michael Graham and secondly in those of Beverton and Holt. All that was needed was to increase the mesh size in the cod ends of the trawls. Thirty years later that view appears innocent. Even in the earliest days the limits were plain – the industrial fishery, *Nephrops* and above all the least management implicit in a trawl fishery for fishes of different sizes.

But the science was carried forward in ICNAF, in ICES and in the IATTC. In the Atlantic the fleets increased in size. When the mesh sizes in the trawl cod ends had reached a least optimum the increment in fishing power generated a new stage of overfishing. The second trend was the appearance of fishing vessels off the shores of countries foreign to them, as distant-water vessels ranged all over the continental shelves; this was a trend that led later to management by the coastal state as a consequence of the Law of the Sea Conference.

13

The second industrialization

The second industrialization of fisheries started in the 1950s and lasted until the 200-mile limits became effective in the early months of 1977. New markets were developed for fish meal and for frozen fish and shrimp. New modifications to gear increased the efficiency of capture. Russian and Japanese fleets expanded their distant-water fisheries greatly, as did the eastern European and Spanish fleets. It was this expansion off the coasts of many nations in the Third World which led to negotiations in the Law of the Sea Conference and to the *de-facto* establishment of 200-mile limits.

The origins

The demand for herring in Germany was high after the First World War, and between 1924 and 1935 a new herring trawl was developed; the headline was raised with high-flying kites, and fish were taken in large quantities. Between 1935 and 1938, the German herring trawlers worked on the Fladen and Dogger grounds in the North Sea. This development in trawling methods was pursued after the Second World War, when high headline trawls and, later midwater trawls were used in many fisheries. Indeed in 1969, West German trawlers started to use very large midwater trawls with sonar to catch capelin on the Grand Banks (Warner 1984).

A second major innovation was the stern trawler used as a factory. The reason for this development in Britain was that, as the distant-water vessels worked further afield with longer voyages, the ice did not last and parts of the catch would be discarded. It originated at Salvesen's, the British whaling company. In March 1954, *Fairtry I* was launched at John Lewis' yard in Aberdeen. She had the whale factory's stern ramp, Baader filleting machines, multiplate freezers from Birdseye and the offal was converted to fish meal. She used a large double trawl. In July 1956, Captain Leo Romyn was fishing the Grand Banks in *Fairtry I* when two Russian *Fairtry*-types appeared, the *Pushkin* and the *Sverdlovsk*; by

Table 13. *Outlets of fish production 1928–83 (10^6 tonnes)*

	1938	1948–52	1953–7	1958–62	1963–7	1968–72	1973–8	1978–83
Fresh	11.1	10.1	12.6	15.9	17.9	18.9	19.1	20.3
Freezing	—	1.2	1.9	3.5	6.0	9.5	12.5	13.2
Curing	5.7	5.7	7.1	7.6	8.2	8.1	7.9	8.1
Canning	1.5	1.9	2.7	3.6	4.7	6.2	9.3	9.7
Meal, oil	1.7	2.0	3.6	8.0	16.2	22.7	19.0	20.0
Misc.	1.0	1.0	1.0	1.0	1.0	1.0	1.0	1.0
Total	21.0	21.9	28.9	39.6	54.0	66.4	68.8	72.3

From Gulland (1983).

1958, twenty-four *Pushkin*-types and twelve of a larger class were fishing the Grand Bank. The Soviet vessels were built by the Howaldtswerke yards in Kiel, from blueprints obtained during Russian negotiations with John Lewis (Warner 1984).

The British had a long history of distant-water fishing, as described in earlier chapters. With the invention of the steam trawlers they explored the coasts of Morocco (1912), Iceland (1900–10), and the Barents Sea (from 1929 onwards). After the Second World War, the side trawlers, or sidewinders, and stern trawlers worked the Iceland grounds, the Barents Sea and the shelf off West Greenland.

The third innovation was the development of market for fish meal and oil. Pelagic fish, such as herring and sardine, have yielded oil for centuries, as in the Swedish herring fishery (1758–1805) or the American menhaden fishery (1870–80). The use of meal for animal food developed more recently, when chicken farmers used it to add essential amino acids to the feed for their birds. The demand for fish meal increased sharply with the market for fast foods. Industrial fisheries developed primarily for pelagic fish, which produced large quantities of fish meal. Table 13 shows the development of various outlets.

During the thirty years since the Second World War, the world catches of fresh fish were doubled to twenty million tonnes, those of frozen fish were augmented by an order of magnitude to thirteen million tonnes and those of meal and oil were also increased by an order of magnitude to twenty million tonnes. Thus half the increment in catches since the Second World War is due to the new markets and methods associated with the second industrialization.

There were other technical innovations, of which two are pre-eminent.

The first was the recording echosounder, which allowed skippers to 'see' the fish beneath him and hence to search for them. Today the signals are presented in colour from a variety of beams, some of which are complex. Sonar, which searches ahead, found a particular use with purse seines and midwater trawls. The second important innovation was the hydraulic power block introduced by Mario Puretic in 1955 in San Diego, which allowed the purse seiners to work on the open sea (because the small boat was no longer needed). For example, the Norwegians moved from their fjords and skerries to the East Icelandic current in winter, one of the roughest places in the world ocean.

The new distant-water fisheries

The main new distant-water fisheries were those of Japan, Russia and Spain. The Spaniards worked across the Atlantic centuries ago, as recounted in earlier chapters, and their pair trawlers still work the Grand Banks where they can. But it was the stern trawler that the Catholic Spaniards used as they steamed south in the Atlantic to augment their intake of fish. The proximate origin of the Russian development has already been described, but it really dated from the time of Lenin, who foresaw that there might be protein shortages in the future, unless augmented from the sea. The Japanese faced the same problem and in the 1920s and 1930s organized exploratory voyages acroᵤ the Pacific that formed the base of the exploitation in the second industrialization (Figure 81).

The Japanese fisheries
(*Kasahara 1961, 1964, 1972;* Yearbooks of Fishery Statistics)

The greatest single component of the second industrialization was the expansion of Japanese fisheries. However, considerable quantities were landed long before, between 1915 and 1940 (Figure 82(*a*)); between 1908 and 1912, 136 trawlers were built in Japan, most of which were electrically powered. Many worked in the East China Sea for sparids. Pair trawls started to work there in 1920 and by the end of that decade, 700 were fishing there with 400 based in China and Taiwan. Later there were 2000 powered draggers working the waters around Japan and in the East China Sea. Danish seines were used for flatfish off the Kurile Is., off Sakhalin, north-east of Korea and off Primorski. The king crab fisheries for canning started off Sakhalin and the northern Kuriles in 1905, and by 1927 eighteen motherships with tangle nets were operating off Kamchatka.

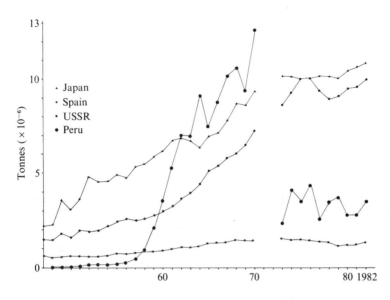

Figure 81. Annual total catches, 1947–82, Spain (■), Japan (▲), USSR (●), Peru (●) (*Yearbooks of Fishery Statistics* for 1947–82).

This fleet migrated to Bristol Bay in Alaska in 1930. Before the Second World War, the tuna fleet comprised 2000 small vessels; fish were caught by pole and line and the vessels did not work far from their home ports. Skipjacks were landed at various bases on Pacific islands.

The main increase in Japanese catches, however, occurred in the last thirty years (Figure 82(*a*)), with a maximum of just over ten million tonnes in 1981 and indeed the growth rate was steady between 1945 and 1973. The trawl fisheries around Japan and in the east China Sea continued, but the real increase came from the broad shelves of the Bering Sea and the Sea of Okhotsk with factory ships and trawlers. Figure 82(*b*) shows the catches of flatfish, predominantly from the North Pacific Ocean; catches of yellowfin sole, arrowtooth halibut, sole, remained more or less steady. About 7000 tonnes year^{-1} of flathead soles and 10000–15000 tonnes year^{-1} of bastard halibut (*Paralichthys olivaceus*) were caught; between 1965 and 1976 about 5000 tonnes of the latter were taken per year. The greatest catches comprised unidentified flatfishes, with an extraordinary peak of more than 500000 tonnes in 1961. It may have been composed largely of yellowfin sole, most of which was used for fish meal.

In Figure 82(*c*) are shown the catches of gadoids (hake and pollack) by

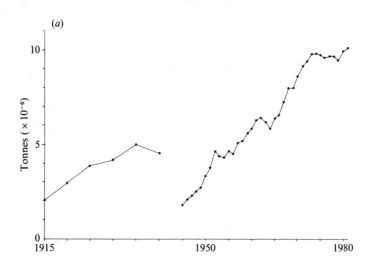

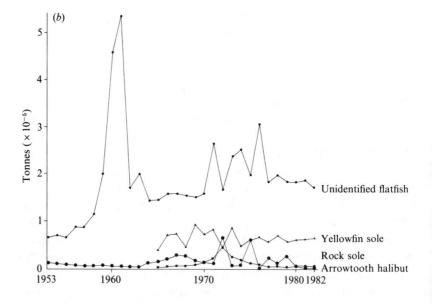

For caption, see p. 241.

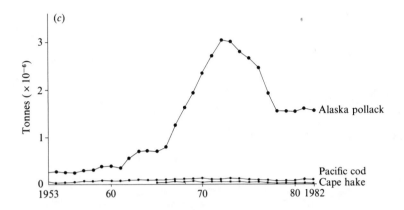

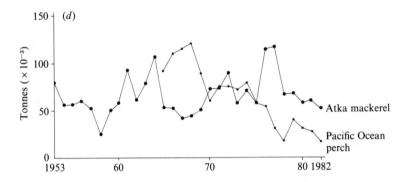

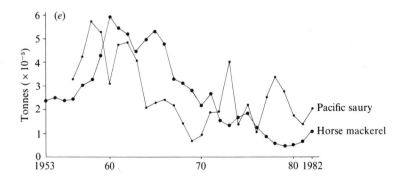

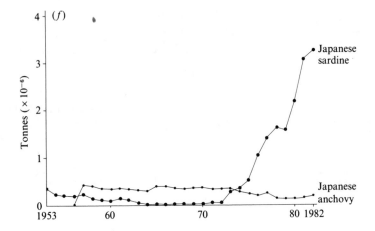

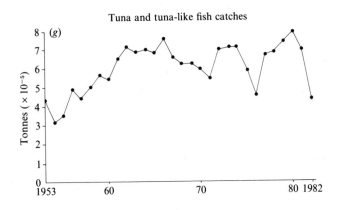

Tuna and tuna-like fish catches

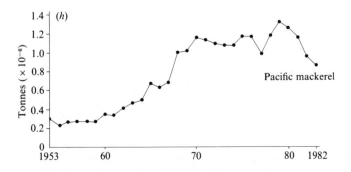

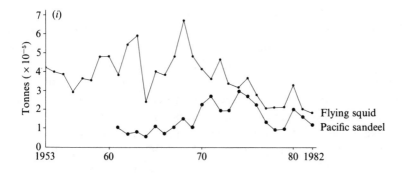

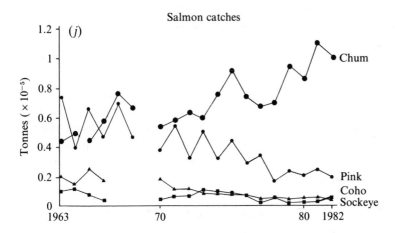

Figure 82. Japanese catches: (*a*) annual marine landings in millions of tonnes from 1915 to 1981; (*b*) unidentified flatfish, yellowfin sole, and arrowtooth halibut in thousands of tonnes from the North Pacific Ocean; (*c*) Alaska pollack and Pacific cod in millions of tonnes from the North Pacific Ocean; catches of Cape hake (*Merluccius paradoxus* and *M. capensis*) are included in this figure; (*d*) Atka mackerel and Pacific Ocean perch in thousands of tonnes from the North Pacific Ocean; (*e*) Pacific saury and jack mackerel in hundreds of thousands of tonnes from the North Pacific Ocean; (*f*) Japanese sardine and Japanese anchovy in millions of tonnes; (*g*) catches of tuna of many species in hundreds of thousands of tonnes from the world ocean; (*h*) catches of Pacific mackerel in millions of tonnes; (*i*) catches of flying squid and Pacific sandeel in hundreds of thousands of tonnes; (*j*) catches of Pacific salmon (*Yearbooks of Fishery Statistics*, vols. 1–42).

the medium trawlers, large trawlers and mothership trawlers in the North Pacific Ocean; a minor quantity was taken by gill nets and long lines. The Pacific cod had been caught before the Second World War by long lines off Kamchatka, Sakhalin, the Kurile Is. and Korea; after that war, trawls were used as well as long lines. A small fishery of 60 000–80 000 tonnes was practised, with trawls for Cape hake (*Merluccius capensis* Cast.; *M. paradoxus* Franca) off South Africa. By far the biggest fishery was that for Alaska pollack with medium, large and mothership trawlers, which peaked at three million tonnes in 1973 and 1974; since then, catches have been reduced by a factor of 3. The major fishing grounds lie off Bristol Bay, in the central Bering Sea, off Kamchatka, Sakhalin, in the Japan Sea and south of Hokkaido (Koslow 1976). Serobaba (1977) suggested that there are four stocks: east Bering Sea, north Bering Sea, west Bering Sea and along the island chain bordering the Sea of Okhotsk. Although much of the catch went to fish meal, a fair proportion ended as frozen surimi (minced fish).

Quite a lot of bream-like fishes were landed in Japan, croakers, drums, rockfish, gurnards and sablefish. The most important species are the Atka mackerel and the Pacific Ocean perch (Figure 92(*d*)); catches of the former have varied between 25 000 and 120 000 tonnes each year. Catches of the Pacific Ocean perch have declined by a factor of about 4 since 1968, when 120 000 tonnes were caught. The Atka mackerel were taken by one-boat purse seines and the perch by trawls.

The group of jack mackerel includes flying fish, mullets, *Decapterus* and yellowtails (much of which is now cultivated). But the largest catches since 1953 have been of jack mackerel taken mainly by one-boat purse seiners and of Pacific saury, caught predominantly by dip nets under strong lights at night. During the late 1950s and early 1960s about one million tonnes of the two species were caught (Figure 82(*e*)). Since then, catches have declined by a factor of 4 or 5 in both species. The catch of saury comprised mostly the annual recruits before first spawning; the main fishery all year round was mainly where the Kuroshio and Oyashio currents mix.

Catches of the Hokkaido herring the Japanese sardine both declined before the Second World War, as noted above. Since 1953, catches of herring (off Hokkaido, the northern Sea of Okhotsk and off Kamchatka) have ranged between 10 000 and 100 000 tonnes year^{-1}; between 1965 and 1979, up to 36 500 tonnes year^{-1} were taken from the south-east Alaska stock. Up to 56 000 tonnes year^{-1} have been caught of round herring (*Etrumeus teres* (de May)). Figure 82(*f*) illustrates trends of catches of the Japanese anchovy and the Japanese sardine. Some 400 000 tonnes year^{-1} of anchovy were taken between 1957 and 1973, after which catches declined. But the most remarkable phenomenon was the recovery of the Japanese sardine around Kyushu and west of Honshu from 1972 onwards; by 1982, annual catches had reached 3.3 million tonnes, about twice the peak catch of the 1930s (see below). Sardines were taken by one-boat purse seines, but the anchovies were caught by both one- and two-boat purse seines.

After the Second World War, the Japanese tuna fishery spread throughout all the subtropical oceans to satisfy the American market for frozen tuna. Figure 82(g) shows the annual catches of tuna between 1953 and 1982; they comprised skipjack caught by pole and line, bluefin, southern bluefin, yellowfin, albacore and bigeye, all of which were taken with pelagic long lines with shiny unbaited hooks. Other tuna-like fishes were caught: Spanish mackerel, frigate mackerel, marlins, sailfish and swordfish. The Japanese tuna fishermen (joined later by their fellows from Korea and Taiwan) ranged across the world ocean between 40° N and 40° S, but the greater proportion of catch was taken in the western Pacific.

Figure 83(h) shows the rising annual catches of the Pacific mackerel (*Scomber japonicus* Houttuyn) taken by one-boat purse seines, but lift nets and hooks and lines have also been used. There is a large purse seine fishery in summer off Hokkaido and a deep purse seine fishery in the Tsushima Strait in winter. More than a million tonnes of mackerel were caught between 1968 and 1980 in the Sea of Japan. Barracuda and hair-tails were taken in relatively small quantities (some tens of thousands of tonnes each year).

There is a minor fishery for Japanese cuttlefish, and separately for long-finned squid and unidentified squid amount to between 20 000 and 115 000 tonnes each year. However the major fishery exploits the flying squid (Figure 82(i)), catches of which varied between 200 000 and 600 000 tonnes each year; they have declined somewhat in recent years. Most of the catch was taken by jigging at night under lights and was used fresh or frozen. The main fishery occurred off Hokkaido, northern Honshu and in the Tsuguru Strait. Figure 83(i) also includes the time series of catches of the Pacific sandeel between 1961 and 1982; they peaked in 1974 at 300 000 tonnes.

The Japanese salmon fisheries started before the Second World War and there was much high seas exploration. Figure 83(j) shows the high seas catches (west of 180° W) of pink, coho, sockeye and of chum salmon; catches of the latter increased steadily but those of the other three species have declined; there are also minor catches of cherry (masu) and chinook.

The fish come from both Asian and North-American rivers, the former predominantly.

The disposal of Japanese catches is displayed in the following table:

	Catch (million tonnes, wet weight)					
	1970	1971	1972	1973	1974	1975
Fresh and frozen	3.01	2.85	3.05	3.25	3.44	3.30
Cured	4.15	4.65	4.76	4.90	4.96	4.93
Canned	0.63	0.75	0.74	0.70	0.72	0.68
Reduction	1.47	1.60	1.61	1.77	1.63	1.55

Yearbooks of Fishery Statistics for 1970–75.

With some prominent exceptions the extraordinary rise in Japanese catches occurred in waters not too far from the home ports. The major exception was the fishery for tuna in the subtropical oceans; but Japanese trawlers did work off south-west Africa, off north-west Africa, off New Zealand, in the Gulf of Aden, off Guyana and in Indonesia. There were two components in the rise of catches nearer Japan, first the catches of groundfish right across the shelves of the North Pacific Ocean and secondly the catches of pelagic fishes in the waters near Japan. The catches of flatfish, Pacific cod, Pacific Ocean perch and Alaska pollack were made in the course of the great exploration of the Bering Sea. This exploration resembles that of the North Sea: as stock density was reduced in the waters around Japan and in the East China Sea, the trawlers moved east into the Bering Sea. The stocks of Alaska pollack had not been exploited before.

The catches of saury, jack mackerel, squid, Japanese sardine, Pacific mackerel and Pacific sandeels all came from waters relatively close to Japan and they appear to be high. Perhaps they should be compared with those on the North American shelf between Cape Hatteras and the Grand Banks, another western-boundary current region. O'Reilly and Busch (1984) have shown that the annual primary production on that shelf is high, up to $350 \text{ g C m}^{-2} \text{ year}^{-1}$ on George's Bank. High productivity associated with the warm core rings inside the Gulf Stream may also occur shorewards of the Kuroshio.

Russian catches

The expansion of Russian catches during the second industrialization is well described in FAO *Yearbooks* from 1953 to 1982. The detail in the statistical tables improves with time, but to show trends the material is lumped for illustration. Then changes will be discussed.

Figure 83(*a*) shows the time series of catches between 1953 and 1982 of Atlantic cod, hakes, Alaska pollack (1962–82), Antarctic krill and Antarctic fish (1977–82) in hundreds of thousands of tonnes. Catches of Atlantic cod reached a peak of 0.986 million tonnes in 1968; they fell to 0.283 million tonnes in 1971, rose again to 0.726 million tonnes in 1974, after which there was a continuous decline until 1982. There were very high catches from the Barents Sea in the late 1960s and from the Atlantic coast of North America in the early 1970s. Since 1974, the Arcto-Norwegian stock has been reduced and, since 1977, the Russians have caught very little cod on the Atlantic coast of North America.

The first hakes taken were the silver and red hake on the eastern seaboard of the United States; peak catches of the silver were 0.35 million tonnes (1965) and of 0.41 million tonnes (1973), but catches of the red were low. Catches of the Cape hake increased steadily to a peak of 0.66 million tonnes in 1972, after which they declined; after 1977, some catches were still being landed and in the period 1978–82 small catches of Benguela hake were taken off Namibia. Senegalese hake were landed in small quantities in the 1970s and off the Argentine, 0.51 million tonnes were taken in one year, 1967. This short review shows how the Russian fleet has had to shift ground from year to year.

But Figure 83 (*a*) also shows the rise in landings of Alaska pollack from the late 1950s and early 1960s to 2.5 miillion tonnes in 1982. This resource, the planktivorous gadoid, is shared by Russians and the Japanese in the Bering Sea. But the Russians also ventured into the Antarctic where in the period 1977–82 they caught about 0.1 million tonnes of fish and up to 0.5 million tonnes of krill; the fleets, therefore, ranged from one end of the Pacific to the other.

Figure 83(*b*) illustrates the Russian catches of three groups, clupeids, jack mackerels and mackerels, between 1953 and 1982. The clupeids include Atlantic herring, Pacific herring, Black Sea anchovy, South African pilchard, sardinellas, Black Sea sprat and Baltic sprat. Catches of Atlantic herring (Newfoundland, Atlanto-Scandian and North Sea herring) peaked in 1964 (0.62 million tonnes), after which they declined to a residual 0.1 million tonnes from Newfoundland; the other two fisheries had been extinguished by recruitment overfishing. Catches of the North Pacific herring remained at around 0.3 million tonnes until they were sharply reduced after 1977. Catches of Black Sea anchovy and of Baltic sprat, local to the Russian fleets, have remained more or less steady at about 0.2–0.3 million tonnes during the whole period 1953–82. South

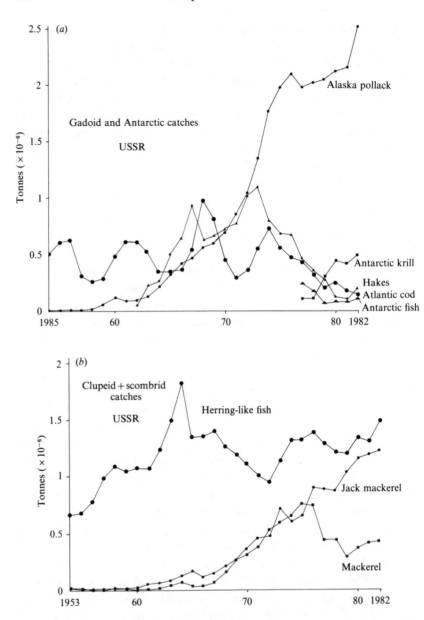

Figure 83. Catches made by the USSR: (*a*) catches of cod, Alaska pollack and of Antarctic resources; (*b*) catches of clupeids, jack mackerel and mackerel, 1953–82. The clupeids include Atlantic herring, Black Sea anchovy, North Pacific herring, sardinellas, Black Sea sprat, Baltic sprat and South African pilchards. The jack mackerels were caught in the North Atlantic, off South Africa, off Namibia and off Chile (*Yearbooks of Fishery Statistics*, vols. 1–42).

African pilchards yielded catches of about 0.1 million tonnes in the early 1960s. Catches of sardinellas remained at about 0.1 million tonnes during the 1960s but reached a peak of 0.57 million tonnes in 1976, after which they declined to about 0.2 million tonnes; most were taken off the coasts of West Africa and north-west Africa. Thus the Russians maintained catches from various parts of the world, excluding the Indian Ocean.

Figure 83(*b*) also shows the increase in catches of jack mackerel, from the Atlantic, the Chilean coast and off the coast of Namibia, from the late 1960s to a peak of 1.2 million tonnes in 1982. The figure also shows the mackerel catches, 0.76 million tonnes, which declined after 1977 as the Russian fleets were excluded from the waters of the European Economic Community (EEC). There are other catches, not illustrated, particularly of capelin which reached high levels (0.61–1.0 million tonnes between 1975 and 1982), mainly from North American waters.

The Russian fleet comprises mainly stern factory trawlers, some of which are very large; they are based on Murmansk, Kaliningrad and Valdivostock. They also work from distant bases, such as Havana. They have been very successful in maintaining catches after 1 January 1977, primarily by arrangement with the countries in whose Exclusive Economic Zones (EEZs) they work. Where they have been excluded, as for example, in EEC waters, they buy much of the mackerel caught from the European trawlers.

Spanish fisheries

Figure 84(*a*) shows the increase in total catches landed in Spain from 1938 to 1969. Initially, catches amounted to 0.4 million tonnes and in the early 1970s landings had reached 1.5 million tonnes, a quantity which declined later to 1.2–1.3 million tonnes (*Yearbooks of Fishery Statistics*). Before 1970 in the FAO *Yearbooks*, Spanish catches were recorded in groups of species (e.g. clupeids), but after that date (as for other countries, catches were recorded in species groups by broad areas. In general the increase in catches dated from the late 1950s and early 1960s and comprised largely landings from stern freezer trawlers.

In Figure 84 total catches are shown and those by regions. Catches from the north-east and north-west Atlantic include cod, together with hake; catches of these species increased by about a factor of 3 between 1953 and 1969. The clupeid catches (anchovy and sardine) came from both Spanish coasts, Atlantic and Mediterranean. The group of redfish-like animals includes Mediterranean fishes (and horse mackerel). The tuna-like fish were mainly (but not entirely) of Atlantic species.

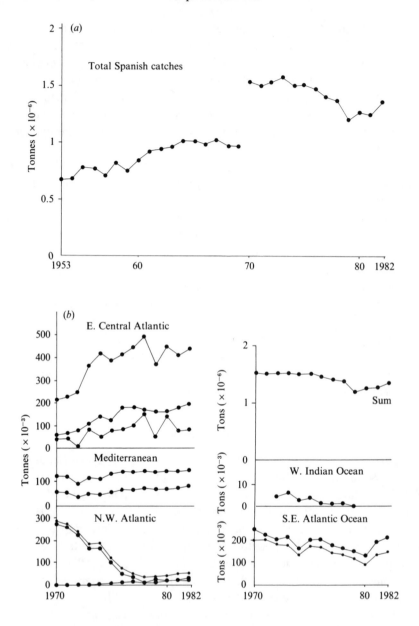

Figure 84. Spanish catches: (*a*) total catches between 1938 and 1969 (FAO records became more complete in 1970); (*b*) catches in the period 1970–82 (*Yearbooks of Fishery Statistics*, vols. 1–42).

Figure 84 also shows catches in the south-east Atlantic, which include those of Cape hake from off South Africa between 1970 and 1982. Hake is a traditional Spanish dish and exploitation of the Cape hake by the Spaniards has continued all through the recent development of this fishery. Catches of mussels from the Rias have reached 100000 tonnes in some years (included in north-east Atlantic catches); they are grown on ropes suspended from rafts in the drowned valleys, which are fed by the upwelling offshore in the summer. There was also a very profitable shrimp fishery off the mouth of the Zambezi in the western Indian Ocean.

The distant-water catches were landed in Vigo in north-west Spain and much of the fish was carried to the growing city of Madrid in freezer trains. In most of Europe and North America, the consumption of fish as a proportion of animal protein has declined since the 1920s but in Spain that proportion has remained high.

The pelagic fisheries

As indicated above, the demand for fish meal after the Second World War led to the fuller development of fisheries for herring, sardines and anchovies. Many of the new fisheries were worked by purse seiners with power blocks, echosounders and sonars. In all parts of the world there was rapid expansion of such fisheries followed by collapse.

North-east Atlantic herring

Figure 85 shows catches in time series of five groups of herring in the north-east Atlantic. In Figure 85(a) are illustrated the trends in catches in North Sea autumn spawning herring. Three spawning groups, Downs, Dogger and Buchan collapsed in succession. Recovery was prevented by the Danish industrial fishery for immature herring. Figure 85(b) shows the time series of catches of the Norwegian spring spawning herring. Until the early 1950s catches were made mainly by drift net, but in the late 1950s the power block allowed the purse seiners to work in the open sea. But the fat herring, the immature herring in the fjords, were also exploited and because between 300000–400000 tonnes were taken each year, the numbers caught were very large. The drift net market as pursued by Norwegians, Russians and Icelanders was for canned or fresh herring, but the purse seiners fished for fish meal (and oil) only. The Norwegian herring stock produces large year classes rather sporadically. The last good one was that of 1950, although there was a fair one in 1959. The com-

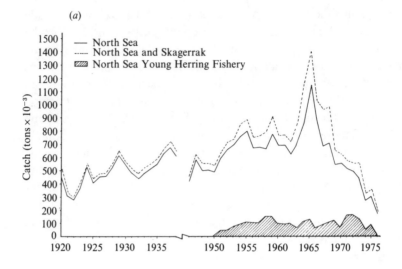

(*a*)

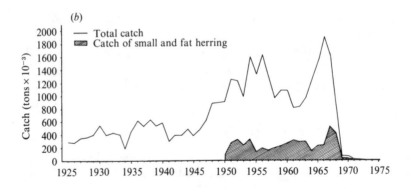

(*b*)

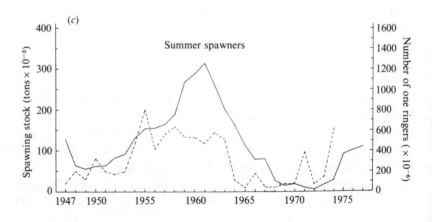

(*c*)

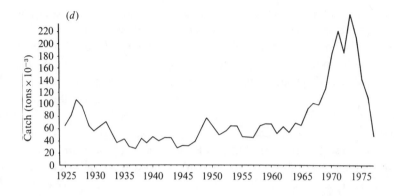

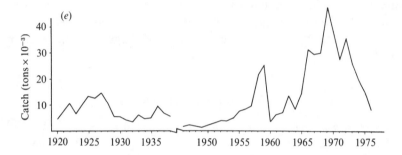

Figure 85. The pelagic fisheries; catches of north-east Atlantic herring: (*a*) North Sea autumn spawning herring, 1920–76; (*b*) Norwegian spring spawning herring, 1925–76 (Schumacher 1980); (*c*) Icelandic summer spawning herring, 1946–80 (Jakobsson 1980); (*d*) west of Scotland autumn spawning herring, 1925–77 (Schumacher 1980); (*e*) Celtic Sea winter spawning herring, 1920–76 (Schumacher 1980). (1 ton = 1.016 tonnes.)

bination of immature fishery and adult fishery (both with purse seines) generated recruitment overfishing.

Figure 85(*c*) shows the time series of catches from the stock of Icelandic summer spawners, which reached a peak of more than 0.1 million tonnes in the early 1960s (Jakobsson 1980). The catches were converted to fish meal. The figure shows two interesting points: the period of high catch from 1961 onward occurred when sonar was introduced in the purse seine fishery. After 1965, a number of minor regulations were introduced, but, in 1972, purse seining was banned and the stock recovered quickly enough for catches to be resumed in 1975, so the stock had suffered from recruitment overfishing.

Figure 85(*d*) and (*e*) illustrates the trends in time of catches of autumn-spawning herring west of Scotland and in the Celtic Sea, respectively. The fishery off the west coast of Scotland was worked by drift nets, but purse seines were introduced in the 1970s; the stock collapsed and fishing was banned in 1976. The Celtic Sea fishery was practised by drift nets in the early years, but in the early 1960s it became a trawl fishery; the stock collapsed in the 1970s and catches were banned in 1976.

To summarize, five stocks of herring in the north-east Atlantic collapsed under the pressure of fishing for fish meal. In the North Sea, in the Norwegian Sea and off Iceland, recruitment overfishing was shown; off the west coast of Scotland and in the Celtic Sea, recruitment overfishing was presumed. In the North Sea and in the Norwegian Sea, catches of immature fish played an important part in generating recruitment overfishing. Purse seiners may catch immature fish and discard them, unrecorded, because the skipper cannot distinguish mature from immature fish on sonar or echosounder.

North American stocks

Figure 86(*a*) shows the catches of north-west Atlantic herring from 1920 to 1970. Until the 1960s, the fish were taken mainly as immatures in weirs on the coasts of Maine and Nova Scotia. In the 1960s, Russian trawlers started to work on the adult stock on George's Bank (in the Gulf of Maine). The peak catch of nearly 1 million tonnes was landed in 1969. Again, it was the combination of fisheries for juveniles and adult fishes which generated recruitment overfishing (Anthony and Waring 1980); the main gear used was the bottom trawl.

The time series of catches of the British Columbian herring is illustrated in Figure 86(*b*). Originally a drift net fishery for dry salt herring, it became a purse seine fishery for fish meal in 1935. The stock collapsed in the middle 1960s and catches were banned, whereupon the stock recovered, having suffered from recruitment overfishing (Hourston 1980). Figure 86(*c*) shows the catches of mackerel between Cape Hatteras and Nova Scotia from 1804 to 1977. From 1850 to 1870 it was a purse seine fishery, and it was in this fishery that the purse seine, in its modern form, was invented. Catches remained below 0.1 million tonnes, but increased sharply with side trawlers, stern trawlers and purse seiners and then collapsed in the early 1970s (Anderson and Paciorkowski 1980). The collapse was due to recruitment overfishing when fishing mortality became high in 1974–7. In Figure 86(*d*) are illustrated the catches of the menhaden fishery off the eastern seaboard of the United States from

1940–77, with purse seines (Schaaf 1980). The collapse in catches was due to recruitment overfishing, but there was an environmental component in the collapse, as frequently happens.

In this group of fisheries, the purse seine in its modern form predominated and the output was fish meal. All suffered from recruitment overfishing to some degree.

Fisheries for sardine and anchovy

Figure 83(*f*) shows the recent rise of the Japanese sardine; catches peaked in the 1930s and the figure shows its recovery in the 70s. Originally it was a purse seine fishery which took a large proportion of immature fish (Kondo 1980). The cause of recovery is not known, but it was the Pacific subpopulation which did so. In Figure 97(*a*) are illustrated the changes in the sardine stocks off South Africa and Namibia, both of which are purse seine fisheries. A peak in sardine catches occurred in 1962 off South Africa and in 1968 off Namibia (Crawford, Shelton and Hutchings 1980). Both collapses were due to recruitment overfishing (Newman and Crawford 1980; Butterworth 1983). An interesting point is that, as sardine catches declined, catches of anchovies increased.

The collapse of the Peruvian anchoveta is shown in Figure 87(*b*). The largest fishery in the world collapsed in 1971–3 due to recruitment overfishing under the adverse environmental conditions of El Niño in 1972–3 (Cushing 1982). Again the catches of sardine tended to replace those of anchoveta after the collapse. In Figure 87(*c*) are shown the time series of catches of the Californian sardine, the Canary sardine, the Chile sardine, the Namibian and the South African sardine. The purse seine fishery for sardines off California was the first major pelagic fishery to collapse under the pressure of fishing, but there was almost certainly an environmental component (Clark and Marr 1955; Marr 1960). This stock was replaced in abundance by the northern anchovy (McCall 1980). The catches of Canary sardine also collapsed and those of the Chile sardine rose to a peak in 1980.

The collapses of the sardine fisheries were followed by that of the anchovy fisheries in four cases, a phenomenon which remains unexplained. Recruitment overfishing was shown off South Africa and Namibia and presumed off Japan. It probably occurred off California and Peru and in both fisheries there was an environmental component in the collapse. In this section, on pelagic fisheries I have attributed collapses to recruitment overfishing where the strict evidence is perhaps absent. The old dichotomy between the effect of fishing and the effect of the environ-

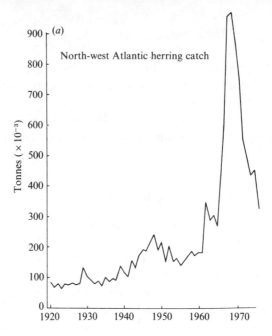

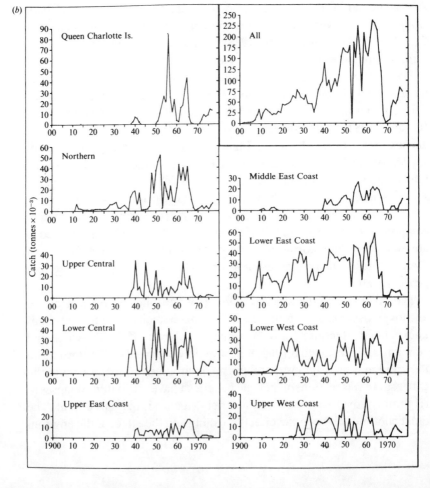

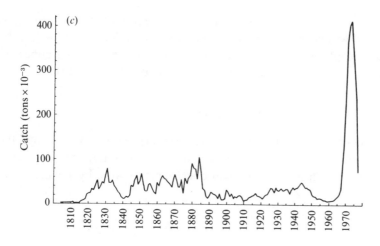

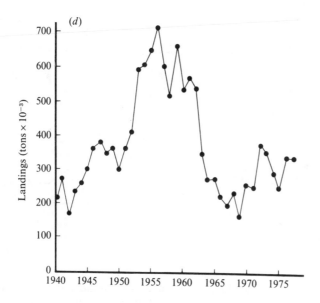

Figure 86. The pelagic fisheries: catches from the North American fisheries. (*a*) North-west Atlantic herring, 1920–70 (Anthony and Waring 1980); (*b*) British Columbia herring, 1902–77 (Hourston 1980); (*c*) north-west Atlantic mackerel, 1804–1977 (Anderson and Paciorkowski 1981) (1 ton = 1.016 tonnes); (*d*) Atlantic menhaden, 1940–77 (Schaaf 1980).

ment may well mislead us. The state of recruitment overfishing is revealed within the variability of recruitment. If death by fishing is the predominant factor, and two or three very poor year classes appear in succession, the stock may still collapse by recruitment overfishing.

Conclusion

The second industrialization resulted from (1) the stern trawler and the demand for frozen fish, and (2) the purse seine and stern trawler and the demand for fish meal. The main developments were by Japanese and

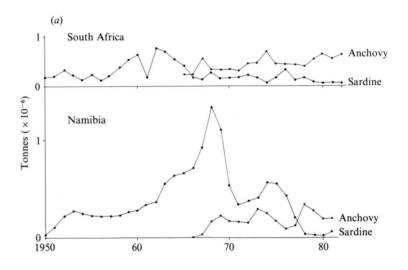

Figure 87. The pelagic fisheries; sardines and anchovies: (*a*) sardines and anchovies off South Africa and Namibia (Butterworth 1983); (*b*) anchovies and sardines off Peru (*Yearbooks of Fishery Statistics* for 1949–81); (*c*) sardines off California, in the Canary Current, off Chile, off South Africa and in Walvis Bay.

Russian fishermen together with the pelagic fishermen throughout the world. But there were many others. Danish industrial fishermen in the North Sea by 1974 landed 2 million tons of Norway pout, sprats and sandeels; they used trawlers with small crews and landed their catches for fish meal. Capelin were exploited off northern Norway and off the Grand Banks for fish meal. British stern trawlers worked in the Barents Sea, as had the earlier sidewinders since 1929, and off Iceland, as had their predecessors from the Suffolk shore in the fifteenth century (see Chapter 3). American tuna purse seiners from San Diego sailed to the coast of Angola. French shrimpers worked off Guyana. The Russians were joined

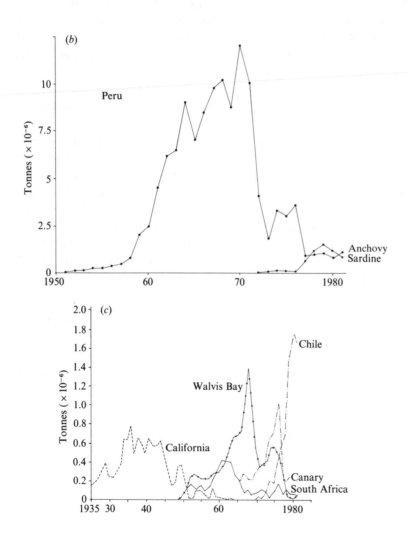

in many of their explorations by Polish, East German, Bulgarian and Romanian stern trawlers. Ghanaian trawlers exploited most of the coast of West Africa between the Gambia and the Congo. Trawlers from Thailand fished off Bangladesh and in the South China Sea. The South Koreans followed the Japanese in their search for tuna. In 1949, at a United Nations Scientific Conference on the Conservation and Utilization of Resources conference at Lake Success, Michael Graham had drawn a chart of the marine unexploited resources, which covered most of the world ocean (Graham 1950). Today, nearly all are fished, most by too many fishermen.

There were two consequences of the second industrialization. Because of the spread of fishing vessels off shores foreign to them, the Law of the Sea Conference developed procedures for the establishment of EEZs. Then in early 1977, many, but not all, nations declared EEZs of 200 miles, *de facto*, but not *de jure* because the Law of the Sea has not yet been agreed or ratified. The second consequence was that the pressure of fishing increased throughout the world, in the world pelagic fisheries, in the demersal fisheries of the North Atlantic, in the Gulf of Thailand and elsewhere.

The first industrialization took place in the North Sea off the eastern seaboard of the United States and off the north-west Pacific coast, small areas in the world ocean. The use of steam power made the trawl more efficient, so much so that the exploitation of fish stocks spread away from home ports and became an international problem. In the second industrialization, the techniques of fishing were improved in many ways and, because the international commissions were not very successful (participation was voluntary), nations wanted to manage the stocks in their EEZs themselves.

14

Fisheries research since 1965

As shown in the last chapter, the second industrialization was well under way during the 1960s. The optimal age at first capture differs in different species, so in a mixed fishery a compromise solution is usually applied. Indeed, by 1970 the effective limit to mesh regulation had often been reached. When fishing effort increased in the 1960s, the demersal stocks in the Atlantic were once again being heavily exploited. This was the first problem that followed the second industrialization. Hence methods of control, in addition to mesh regulation, were needed. Closed areas and closed seasons in waters, then international, might have been difficult to agree and enforce between the many nations involved. Of the many possible controls, only catch and effort quotas were politically feasible, although the latter might not have been successful. The quality of effort statistics was then so poor as to be unusable and so the only management measure was the catch quota, set anew each year; a quota fixed for years was never considered. As will be described below, a method was devised in 1965 from which annual catch quotas (total allowable catches, TACs) could be readily estimated.

The second problem that followed the second industrialization was that of recruitment overfishing. Many pelagic stocks collapsed, as shown in the last chapter – north-east Atlantic herring, British Columbian herring, Peruvian anchoveta and many others; too many immature fish were caught in some fisheries by purse seines. A particular difficulty is the effect of fishing when poor year classes succeed each other due to a persistence of environmental factors. The anchoveta stock failed off Peru because three poor year classes (two associated with El Niño) succeeded each other when fishing was intense; recruitment had survived two earlier El Niños without failure. The apparent dichotomy between the effect of fishing and that of the environment is misleading; with hindsight, Peruvian managers might have been provided with a procedure which told them what to do when the succession of three year classes appeared (Sissenwine and Shepherd 1987). As will be shown below, the

259

practical problem of warning managers of recruitment overfishing has been approached and there is at least the hope that the dramatic collapses of stocks through that cause will not recur.

There is a further need – to study more fully the factors that affect the generation of recruitment. The magnitude of recruitment may well be determined by about six or nine months after hatching and so an early but preliminary forecast is possible. But there are more important needs: first, to describe the dependence of recruitment on parent stock more fully and, secondly, to establish the mechanisms behind the natural changes which may be quite large and the competition between stocks (if they exist). Was the large 1904 year class of the Atlanto-Scandian herring to be regarded as an act of God or should we try to elicit the underlying causes of such prominent events? The exchange between sardine and anchovy (Troadec, Clark and Gulland 1980) under exploitation is a consequence of recruitment overfishing. Anchovies succeeded sardines off California, off South Africa and off Namibia and sardines succeeded anchovies off Peru. All exchanges occurred under heavy exploitation and under such conditions the effects of natural changes may become amplified.

A third problem is that of multispecies management, a problem which existed before the second industrialization. Any fishery comprises more than one species, three in the Barents Sea, twelve in the Irish Sea, fifty in the Gulf of Thailand and 1200 off north-west Australia, all of which are trawl fisheries. In the simplest case, the mixed fishery is concerned only with differences in fishing tactics, gear and catchabilities. Of more interest are the possible biological links: what happens to the prey of an exploited stock? and how is a fishery to be managed which exploits both predator and prey? There are three parts to the problem: the allocation of catchabilities in a mixed fishery; the examination of gut contents; and the establishment of the competitive links concealed in the stock/recruitment relationships of the many species.

Fisheries research has advanced in the following directions: (*a*) development of the catch equation, (*b*) development of the production model, (*c*) the dependence of recruitment on parent stock, (*d*) the multispecies problem.

The development of the catch equation

Baranov (1918) formulated the catch equation, $C = F\bar{N}$, where C is catch in numbers in a year, F is the instantaneous coefficient of fishing mortality and $\bar{N}$ the average stock in numbers in that year ($\bar{N} = N_0(1-\exp(-Z))/Z$, where N_0 is the initial number in that year and Z is the instantaneous coefficient of total mortality). Thompson and Bell (1934), Ricker (1944), and Beverton and Holt (1957) used it. The formulation can equally well be applied to the catch in numbers in an age group or to all age groups lumped and events in a year class can be described in a string of successive equations. Jones (1961) and Murphy (1965) proposed solutions but the present system was devised by Gulland (1965) by making educated guesses of fishing and of natural mortalities (M). He named this method *virtual population analysis* because the basic data were the catches in numbers at age in the year class, or the virtual population (Fry 1949). Pope (1972) introduced a simplification he called *cohort analysis*. Usage has confused the two names; I shall call both *sequential analysis*. In the ith year of the year class

$$C_i = F_i \bar{N}_i = F_i N_i \, (\frac{1-\exp(-Z_i)}{Z_i})$$

Then $\qquad\qquad N_i = C_i Z_i / F_i (1 - \exp(- Z_i));$

recall that $\qquad\qquad Z_i = F_i + M_i,$

$$N_{i+1} = (\exp(- Z_i) \, C_i Z_i \, / \, F_i \, (1 - \exp(- Z_i))$$

$$C_i / N_{i+1} = F_i (1 - \exp(-Z_i)) \exp Z_i / Z_i$$

$$= F_i \, (\exp(Z_i - 1)/Z_i$$

where M_i is assumed unknown; given C_i, M_i and N_{i+1}, determine F_i and calculate N_i.

Pope (1972) simplified and clarified the method:

$$N_i = \exp(M_i/2) \times C_i + \exp M_i N_{i+1}.$$

The calculation starts with the last age of the year class and proceeds back to the first. The guessed fishing mortality is an educated one which improves as the data array augments. Successive catches at age from older to younger age groups in the year class reduce the error in fishing mortality; Pope (1972) showed that it is reduced to low levels within three years at levels of fishing mortality which are not too high. Hence the numbers in the abundant age groups in the year class (including the recruitment) can be well estimated. The whole procedure can still be biased if natural mortality is poorly estimated or if it changed with age (Agger, Boetius and Lassen 1973). Hence, it is desirable to check the relative abundance between years with groundfish surveys and egg surveys from time to time. The system works reasonably well with long and stable age distributions. Sometimes it is desirable to separate the fishing mortality exerted in a year from the selectivity pattern of fishing mortality with age (Shepherd and Stevens 1983).

The great advantage of sequential analysis is that the estimates of fishing mortality and of average stock in numbers become available by age and by year for decades, independently of effort data. In earlier years total mortality was estimated from the annual age distribution in stock density and there were a few observations of fishing mortality estimated from tagging experiments. In contrast, sequential analysis has provided much more information (admittedly on the assumption that natural mortality is well estimated). One of the consequences of recent work by the Multispecies Working Group of ICES was that it was found that earlier estimates of natural mortality of adults had proved to be reasonably good. The dependence of fishing mortality upon fishing effort can be established. One of the consequences of the relationships between fishing mortality and fishing effort was that the catchability coefficient, q, was shown to be inversely related to abundance in some stocks (king crabs, Rothschild 1970; Norwegian herring, Ulltang 1976; Arcto-Norwegian cod, Pope and Garrod 1975; Californian sardine, McCall 1976). The question arises whether this phenomenon is general or not, or whether it arises amongst species, such as the North Sea plaice, that do not shoal.

With the von Bertalannfy growth equation, differences in length can be expressed as intervals in time. Pope's (1972) approximate form of sequential analysis makes use of catches in numbers at age at yearly intervals. Jones (1974, 1981) used this formulation for yearly intervals so that length (L) frequencies could be used in place of age distributions. Then the sequential analysis can proceed in the usual way provided that estimates of L_{∞} and (M/K) (see p. 215) are available. Pauly and David (1980) started with Gulland's (1965) form of sequential analysis and expressed the catch equation in terms of any time interval rather than in years ($C_i = F_i \bar{N}_{i+\Delta i}$; estimates of both K and L_{∞} are needed); Shepherd (1987) has used a similar matching method to separate the length modes objectively, provided that length compositions from research vessels are used which include the smaller fish. With the growth parameters determined in some way, the possibility is raised for estimating yield per recruit (Jones 1981) and catch forecasts directly from the length distributions. Such methods are also very useful in those stocks for which age determinations do not exist, as, for example, on many occasions in tropical and subtropical waters. The step to the use of sequential analysis was considerable and today the method is widely used. But we should recall that the information is that of numbers caught at age and that the estimates of fishing and natural mortality can vary in quality.

In recent years, the concept of Status Quo Catch (SQC) has been introduced (Pope 1982); that is, the estimated forecast catch with fishing mortality constant, or nearly so. Simple algorithms have been devised for this purpose (Shepherd 1984). Indices of recruitment can be used in the system, together with variation in stock size, a considerable simplification, which is robust:

$$Y_{(n+1)} = \frac{\overline{F}_{(n+1)}}{F_n} \times Yn \exp (G - Z_n) + pr_{(n+1)},$$

where $\overline{F}$ is the (catch-biomass) ratio, r is the estimator of recruitment; p is the coefficient of the recruitment term (Anon. 1986b). The point is that differences in TACs due to differences in recruitment can be estimated in a simple manner in fisheries which are not well known and in which fishing is relatively stable and, of course, for the well-known stocks.

Perhaps the most important consequence of sequential analysis was its use in the Atlantic for the development of catch quotas or TACs. Because there is an interval between data analysis and action, the catch next year has to be predicted from the stock last year. The estimate of stock last year is potentially biased, but as the data array improves with the years, such biases may become less important as the years roll by. Doubleday (1979), with a sensitivity analysis, and Pope (1982), with a simulation, suggested that the error could be fairly low, *10% under favourable conditions* in the year of estimation, but of course greater two years later. Under poor conditions, the errors of estimation are considerably greater. Brander (1985) has shown that it can be reduced if the TAC comes into force in the year after estimation. Such an estimate of error is of considerable value to managers, provided they know that they might have to work sometimes on the lower bound. The development of the multispecies model in ICES has suggested that it might become possible to consider quotas of a term longer than annual, as additional to the present system of annual quotas.

The catch predictions are, of course, made with respect to a management objective. There are a number of potential objectives, but in practice it has been the maximum sustainable yield (MSY), or something less than that (in terms of fishing mortality). Where stock remain overexploited, the first objective would have been to reduce fishing effort towards that at MSY. Subsequently, it is desirable to use a level of effort less than that which crudely produces MSY to prevent an overshoot due to error in the system. $F_{0.1}$ is a fishing mortality (where the slope of the yield curve against fishing mortality is one tenth of that at the origin) which approximates to a crude economic assessment across the exchanges

(Gulland and Boerema 1973) (but see Chapter 15). Where such assessments are properly feasible (within a national system), a maximum economic yield can be obtained directly (Clark 1985). As noted earlier the phrase maximum sustainable yield has been written into certain treaties and so there is a general meaning as well as the technical one.

The development of the production model

In the Schaefer model (Schaefer 1954, 1957), the curve of surplus yield against stock or against fishing effort at equilibrium is a parabola and the dependence of catch per unit of effort against effort is assumed to be linear. Schaefer and Beverton (1963) noted that, in some stocks at least, the yield curve was skewed towards the origin, as indeed Beverton and Holt (1957) had shown. Pella and Tomlinson (1969) generalized the Schaefer model so that the exponent, m, of the yield curve can take any value rather than be constrained to 2, as in the parabola. If $m < 2$, the yield curve is skewed towards the origin and the dependence of catch per effort on effort forms a concave inverse relationship (as observed, for example, by Thompson 1952). Observations tend to be grouped to the right of the maximum because an equilibrium was not reached during the period of observation. Walter (1986) has proposed a method of obtaining equilibrium yields with a graphical procedure.

Deriso (1980) has developed a delayed-difference form of production model in which the annual changes in recruitment are taken into account and expressed as 'annual surplus productions'. Deriso and Quinn (1983) have combined this method with sequential analysis and have applied it to the stock of Pacific halibut. Deriso (1985), with the same methods, has demonstrated a sharp density dependence in the growth of adult Pacific halibut and a curve of dependence of recruitment against parent stock shaped like a dome.

As noted above, the catchability quotient, q, increases with decreased abundance for certain stocks of fish that shoal. If production models are used on observations on such stocks, MSY will be overestimated if the trend in catchability is not detected. So far, such trends have been detected from estimates of fishing mortality derived from sequential analysis. A virtue of the production models has been that safe results could be obtained with simple observations, and recruitment overfishing should be avoided. But it now seems clear that the method needs support from more developed observations at least for the shoaling fishes.

The dependence of recruitment on parent stock

The models of Ricker (1954) and Beverton and Holt (1957) described the dependence of recruitment on parent stock. Since then, many stocks, particularly those of pelagic fishes, have collapsed by recruitment over-fishing. This is the condition where recruitment is reduced under high exploitation.

The variability of recruitment is high, from a factor of 3, to one or two orders of magnitude. Cushing and Harris (1973) fitted Ricker curves to the data for a number of stocks (Figure 88); the error lines indicate the limits to the fitted curves. At first sight, this was a reasonable solution but it depended on the use of the Ricker curve only. Elliott (1985) has described the dependence of recruitment on the parent stock of brown trout in Black Brows Beck, in Cumbria, in northern England. For a period of twenty years, eggs, alevins, later larvae and juveniles were sampled. Figure 89 shows the dependence of numbers of juveniles on numbers of parent eggs. Further, the later mortality is inversely related to the earlier, which implies a density-dependent control in the later stage. It is a somewhat special case in a microcosm. The dependence of adult recruits upon parent stock is much more variable, perhaps because there is an immigration of adult recruits from other streams.

Shepherd and Cushing (1980) developed a model based on the density dependence of growth in larval and early juvenile stages:

$$N_1 = AN_0/[1+(1-A)N_0/k],$$

where N_0 is the number of eggs, N_1 is the number of recruits, k is a constant related to the food available and A is the time to grow through a critical period if food were superabundant. If the period of density-dependent growth was long, because of food lack, cumulative mortality during that period would be high and vice versa; it is the cumulative mortality which is density dependent, not the coefficient. The model comprises three trophic levels in which the growth and mortality of larval and/or juvenile fish are regulated by both the number of predators and the quantity of food. The slope at the origin is

$$A = \exp\{-\frac{\mu}{G_{max}} \times \ln \frac{W_1}{W_0}\} = \exp(-\mu T_0)$$

where W_0 and W_1 are initial and final weights during the critical period, μ is the instantaneous predatory mortality rate, G_{max} is the maximum

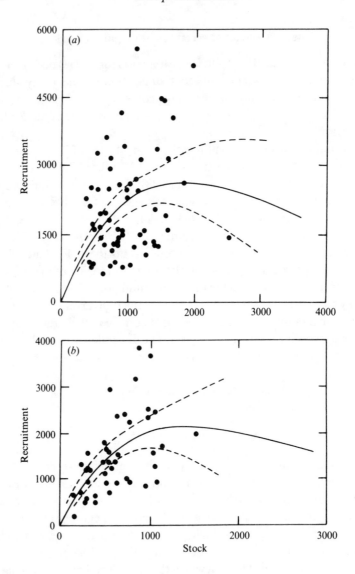

Figure 88. The dependence of recruitment upon parent stock fitted with Ricker curves; the broken lines show the standard error of the fitted line. (*a*) Karluk river sockeye salmon; (*b*) Skeena river sockeye salmon (Cushing and Harris 1973).

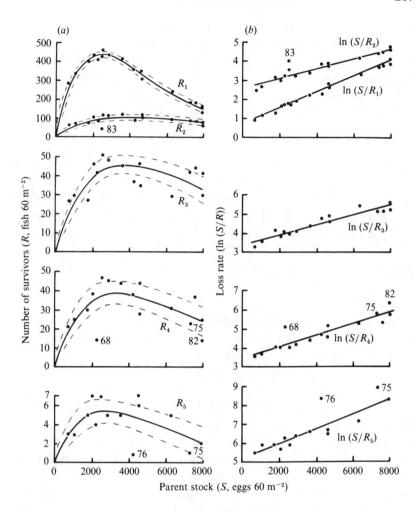

Figure 89. The dependence of recruitment (R) upon parent stock (S) of migratory trout in Black Brows Beck, in Cumbria, in north-west England; a Ricker curve was used and the broken line indicates the 95% confidence limits to the curve. The spread about the line increases from the age of the early parr (May to June) to the age of recruitment (Elliott 1985). The five stages of recruitment are as follows: R_1 0+ parr in May and early June; R_2 0+ parr in August and early September; R_3 1+ parr in May and early June; R_4 1+ parr in August and early September; R_5 females spawning in November and December.

growth rate, and T_0 is the critical period in superabundant food. Another point made by Shepherd and Cushing was that the degree of density dependence in fish stocks was potentially high because many stocks in the north-east Atlantic survive at steady recruitment, where fishing mortality is two to five times greater than natural mortality (Figure 90).

Shepherd (1982) also proposed another convenient model:

$$R = aB/[1+(B/K)^\beta],$$

where R is recruitment in numbers, B is spawning stock biomass, K is the threshold biomass, the magnitude of biomass above which density-dependent effects predominate, and β expresses the degree of compensation. The last of these is probably low for herring-like fishes and high for cod-like fishes; Figure 91 shows the application of Shepherd's equation to the North Sea stock of herring. The curve is fitted with an estimate of the slope at the origin and an estimate of the compensatory coefficient and hence K; indeed, with the judicious use of the exponent, the curve can be drawn to the data directly, without statistical estimation.

There are two safeguard procedures. A model of stock and recruitment can be combined with a conventional yield-per-recruit calculation (Beverton and Holt 1957; Cushing 1973; Shepherd 1982) to give the curve of yield against fishing mortality (Figure 92 was obtained using Shepherd's formulation). Such a procedure might prevent recruitment overfishing. A second method is due to Sissenwine and Shepherd (1987): they propose that the median recruitment per unit stock (with equal numbers of observations on either side of the line of recruitment per stock) be inverted and entered on the curve of spawning stock biomass per recruit against fishing mortality. Then this value of stock per mean recruit indicates a level of fishing mortality which is probably sustainable. With care such procedures should prevent the recurrence of recruitment overfishing as changes take place.

There is a more general requirement, to study the generation of recruitment for forecast in management. Various preliminary attempts have been made, for example the dependence of the recruitment of Pacific mackerel, northern anchovy and Californian sardine on various indices of upwelling (Bakun and Parrish 1980). Another example is the relation between cod recruitment in the North Sea and the delay in the production of *Calanus* in the eastern North Sea (Cushing 1984). What is needed, however, is a more intense study of the trends in growth and mortality of a year class as it develops with respect to the succession of foods and predators. It is really the application of the usual methods of population analysis to the animals in the plankton, a discipline which is much over-

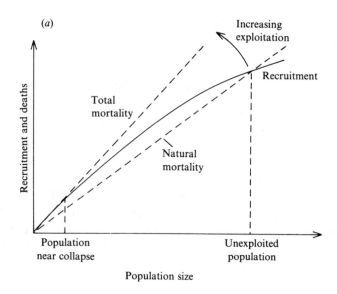

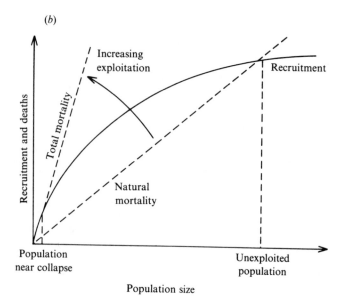

Figure 90. The evidence for strong density dependence in exploited fish stocks. The lower sloping broken line shows the natural mortality and the upper the greatest total mortality: (*a*) weak density dependence (*F/M* = 1); (*b*) strong density dependence (*F/M* >3) (Shepherd and Cushing 1980).

due. As larval fish and juveniles can be aged (Brothers *et al.* 1976), the project should be feasible, if expensive.

The multispecies problem

There are not many fisheries that comprise only one species, but in tropical and subtropical seas outside the upwelling areas there are large numbers of species in the trawl fisheries. The mixed catch arises from a true array of species on the ground or from a tactical search during the voyage; the mix is optimized for different markets, and skippers go to different grounds for different species. Further, there may be a target species and a by-catch: shrimp is the high-priced target and the by-catch comprises species which may be discarded. In any mixed fishery with the same fishing effort, the catchabilities will differ between stocks. So, in a North Sea fishery where there may be half a dozen species caught in a trawl haul, stock densities and catches-at-age may vary somewhat for market reasons.

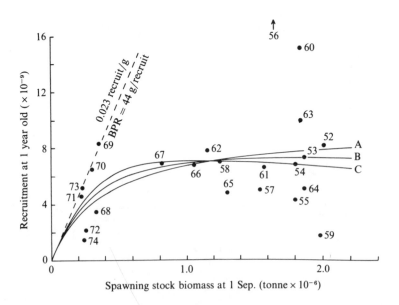

Figure 91. The dependence of recruitment on parent stock in the North Sea herring, fitted with Shepherd's versatile curve; it is fitted with an estimate of the slope at the origin and with three values of the coefficient of density dependence (Shepherd 1982).

There are biological links between species; cod eat haddock and whales eat krill, and both pairs are exploited by man. There are three relationships of potential importance: (*a*) estimates of the predation on pre-recruits will yield estimates of their natural mortality; (*b*) estimates of recruitment from which the effects of predation have been removed might illuminate the stock/recruitment relationship; (*c*) the exploitation of predator and prey at the same time is a complex problem because the predator competes with the fishermen who catch the prey. Further, competition between species is a field unexplored in fisheries research and the mechanisms remain unknown; sardines succeeded anchovies under heavy exploitation (Troadec *et al.* 1980) in the major upwelling areas and on the shelf north of Cape Hatteras, *Ammodytes* succeeded herring which had been heavily exploited (Sherman *et al.* 1981).

The multispecies problem is that of managing many species together rather than one at a time. Beverton and Holt (1957) considered that their model applied to North Sea species as a whole – plaice, cod, haddock, sole etc.; of course the problem was simpler then because the output of

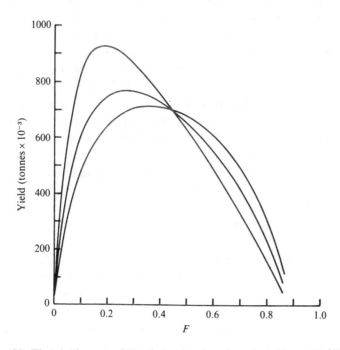

Figure 92. The yield curve of North Sea herring, based on Figure 91 (Shepherd 1982), in which yield per recruit is linked to Shepherd's stock–recruitment relationship. For *F*, see p. 213.

the model was a single age at first capture, or mesh size. When catch quotas were introduced, the management problem changed its nature. For example, the quantities eaten of marketable size become of obvious importance. Ursin (1982), using Daan's (1973, 1978) material on the gut contents of cod, was able to show with a multispecies model that the maximum sustainable yield of North Sea cod was less but at a higher fishing mortality, if the quantities of cod and haddock eaten by the cod were taken into account (Figure 93).

An array of single-species assessments leads naturally to multispecies assessment. There have been various attempts to approach the problem. Pope (1976) combined single-species assessments in multidimensional form. For the North Sea, Andersen and Ursin (1977) introduced a model which found no application perhaps because it was too complex, but included in it were the first formulations of predatory mortality. One consequence of Daan's (1973) study of cod gut contents and of the early multispecies models was that the International Council for the Exploration of the Sea established the 'Year of the Stomach' when a thorough survey was made of the food of commercial species in quantitative terms. Again, as a consequence, Pope (1979) applied the methods of sequential analysis to the numbers of food organisms eaten at age; this device simplified the problem to some degree. Indeed, a multispecies predation model on the same lines was developed by Sparre (1980).

There are many published multispecies models (Brown *et al.* 1975; Brander 1977; Helgason and Gislason 1979; Pauly 1979; Sparre 1980; Larkin and Gazey 1982, amongst others). Shepherd's (1988) model is now being used for the analysis of North Sea material, linked to that of Sparre. It is a yield per recruit model which includes predatory mortalities coupled with Shepherd's (1982) stock/recruitment relationship, for each of eight species in five mixed fisheries. The species preference is given for size of prey (say $0.01 \times$ predator weight, with a variance). A predatory mortality was obtained from a predatory coefficient linked to the species preference (but there is no element of satiation). The working system is constructed in a sequential series.

Preliminary results of these approaches may be found in the reports of the Multispecies Working Group of the International Council for the Exploration of the Sea (Anon. 1986*a*). The most important result was that the natural mortalities of young fishes were estimated and they were considerably higher than the earlier values (which were, as often as not, guessed). The effect was to increase the magnitudes of recruitment and hence the stock biomasses. The average natural mortality of the year class was higher, with the consequence that the curves of yield per recruit

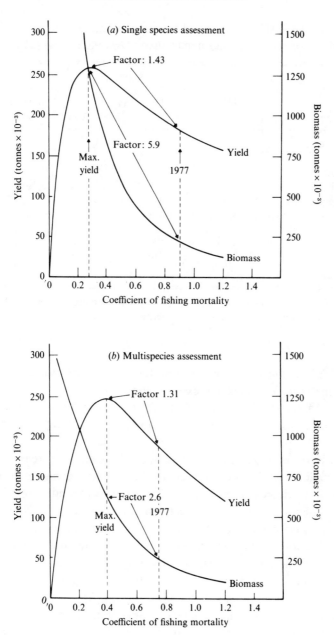

Figure 93. The curve of yield against fishing effort for the North Sea cod: (*a*) single-species assessment; (*b*) multispecies assessment (Ursin 1982). The factor is the ratio between fishing mortality at the maximum yield to that observed in 1977.

Table 14. *Changes in yield expected with decrement of fishing mortality of*
40% for nine species in six North Sea fisheries (Anon 1986a)

	Species							
Fisheries	Cod	Whit-ing	Saithe	Had-dock	Herr-ing	Sprat	Norway pout	Sand-eel
Roundfish human consumption	+	+	S	S	+	S	+	S
Industrial demersal	(−)	(+)	S	S	(+)	S	+	S
Industrial pelagic	S	S	S	S	−	(+)	S	S
Herring	S	S	S	S	(+)	S	S	S
Saithe	+ +	(−)	(−)	S	S	S	+	(−)
Mackerel	S	S	S	S	S	S	S	S

+ +, +, (+), (−) indicate changes expected with an increase (or decrease) in fishing mortality from the present level. S, steady, or no change.

against fishing mortality tended to flatten. This is also true of the yield
curves in the region of present exploitation.

Table 14 shows the changes that might be expected in each of nine
species in six fisheries with a decrement of relative fishing mortalities of
40%.

None of the nine species is affected by an increase (or decrease) in fishing mortality in the North Sea mackerel fishery. An increase in the saithe
fishery would decrease yields of whiting, saithe and sandeel but the cod
yield would benefit. The only effect of an increase in the herring fishery
would increase catches of that species. An increase in the industrial
pelagic fishery would augment sprat catches and diminish those of
herring. An increment of fishing mortality in the industrial demersal
fishery would generate increases in whiting, herring and Norway pout
catches, and a decrement in the cod fishery. In fishery for human consumption of roundfish, all catches would increase, except sprat and sandeel. Thus, it is now possible to quantify Gulland's (1982) conclusion that
the top predator, cod, should be exploited more fully to benefit the other
stocks.

The large-scale changes which are of such interest do not yet come
within the scope of the ICES Multispecies Working Group model. The
reason for this is that recruitment is probably determined at an age below

that at which the model starts to work. Information is needed on the stomach contents of predators in the early stages of the life history and from year to year, but of course the predators must first be identified, which might be difficult. Only then will the problems of recruitment generation be resolved.

But the most important consequence of the model is that the relative virtues of different management objectives in the North Sea have become much less than in traditional single-species assessments. There are three common objectives – greatest yield, greatest value and maximal employment. A reduction in fishing effort of about 40% would produce no pronounced increase in yield, except for herring in the industrial pelagic fishery. Some fishermen would lose their jobs, but there would be economic benefits (higher earnings per day at sea). Hence, within this limited framework of a 40% reduction in effort, the rational exploitation has become the balance between gain in profitability and loss in jobs.

A third consequence is that it has become possible to consider changes in a longer term, say five years or so. For example, would it be desirable to increase the fishing effort in the fishery for human consumption in order to increase catches of herring, haddock and whiting? This raises the question of an active management in which the stocks are manipulated to our advantage (Smith and Walters 1981). In a more limited sense it might become possible to manage stocks in concert so that industry can look ahead.

The fishery in the North Sea is a special one in that the large industrial fishery makes it necessary to estimate natural mortality properly from a very early age. In the seas exploited by Scandinavian countries, the age at first capture is high and the greatest yield remains a reasonable objective for management; but it remains true that the natural mortality of the young fish should be properly estimated from gut content studies. It is also true of the Pacific halibut, where Deriso's (1980) method gives an annual surplus production matched to the annual yield; if less, fishing must be restrained and, if more, fishermen have some freedom to catch fish. The annual adjustment recalls a similar approach by Thompson half a century ago, using catch per unit of effort as an index of stock density (see p. 198).

There are two other problems, the science of tuna-like fishes and the science of the Pacific salmon. In the eastern tropical Pacific, the yellowfin stocks have not yet been defined precisely enough for the purposes of management (Royce 1964; Suzuki, Tomlinson and Honma 1978). In the Atlantic there are analogous problems, apart from the bluefin, which has been overexploited. The problems of the Pacific salmon are of a similar character. The fish return to their native streams to spawn, but each stream is one of many within a river system. There are a number of river systems between Alaska, northern Japan and northern California and the juveniles mix across the Pacific where some eastern stocks are exploited west of the International Date Line.

Conclusion

In the last twenty years the course of fisheries research has changed dramatically. Sequential analysis in its different forms altered the nature and practice of management profoundly from the simple apparatus of mesh size, minimum landing size, closed area, closed season and so on to the more complex business of catch quotas. In some ways management became harder, as is described in Chapter 15, because of the variability of recruitment. The second development in fisheries research was the use of multispecies models based on stomach content data. As a consequence the management objective in the North Sea, at least, was reduced in some cases to the conflict between gain in profitability and loss in jobs. Then the future in the North Sea holds the promise of active management in which changes in stock quantities might be arranged to fit the needs of industry. Elsewhere, at the moment, the yield somewhat less than the greatest remains a reasonable objective (provided that the natural mortality of young fish is well estimated, or that only adult natural mortality is used).

15

Institutions since 1977

In the early months of 1977, many countries declared Exclusive Economic Zones (EEZs) of 200 miles from baselines (which corresponded roughly to the old 3 mile limit from headland to headland). Except for those concerned with tuna, the international commissions lost their management function to the coastal state. The tuna is a 'highly migratory species' (in the phraseology of the Law of the Sea) and so the Inter-American Tropical Tuna Commission and the International Atlantic Tropical Tuna Commission retained their functions in management. The other commissions survived because they became centres of scientific advice to coastal states.

Nations had adhered to the commissions voluntarily and action was limited to the most that could be agreed, and often limited to the least action. It was then the great hope of the Law of the Sea Conference that a coastal state would have the power to take the needed action. This chapter records the attempts made so to do, in the United States, in Canada, in Scandinavia and in the European Economic Community. The account is restricted to these regions because there the brush of science with action is well recorded. But before we proceed, we should examine the part played by economics in the management of fish stocks.

Economics in fisheries management

When Petersen (1894) identified growth overfishing in the Skagerak plaice stock, he thought that fishermen would solve the problem by seeking larger fish. They did not succeed because most had already been caught. Graham (1943) described the history of three fisheries: the trawl fisheries in the North Sea, that for the Pacific halibut in the north-east Pacific and that in the Kavirondo Gulf of Lake Victoria, and he stated his 'Great Law of Fishing':

Fisheries that are unlimited become unprofitable,

277

or fisheries that are unlimited become inefficient; he concluded that the limitation of fishing effort would restore profit to the fishery.

Fishermen fish for money, and the role of economics in management should be made explicit. Gordon (1954) formulated the problem in a Schaefer-like model (see p. 219), where revenue replaced yield as a parabolic function of fishing effort and cost was expressed as a linear function of effort. Between 1954 and 1977, a small literature developed on the subject and in ICNAF there was considerable discussion on the part that might be played by economists in international management. No progress was made in the international fisheries because transfer across the exchanges cannot be easily expressed; indeed that was why Gulland and Boerema (1973) devised the $F_{0.1}$ method (see p. 219).

Clark (1985) described economic models based on (biomass $\times$ value) minus (fishing effort $\times$ cost), as a function of discount rate (the rate at which present money is lost in the future). Thus, the traditional biological parameters are linked to the economic. Revenue is given as a parabolic function of biomass and total cost as an inverse linear function of biomass (Figure 94); the maximal economic yield is given as the tangent to the parabola parallel to the cost curve. Thus the greatest revenue above cost is obtained. These processes were formulated for a sole owner, the ultimate form of limited entry; then revenue was maximized as a function of discount rate. Clark showed that the discount rate is equivalent to the marginal productivity of the resource. Thus, if the discount rate is high, there is every reason to take the whole stock quickly. If the discount rate is zero, all time periods are the same, so the maximum economic yield may be taken steadily. The problem is to set the discount rate properly for quite long periods of time. An interesting model was that of Clark and Kirkwood (1979) for the Gulf of Carpentaria prawn fishery. During the fishing season, the biomass declines and economic criteria were used to decide when in the year to close it.

Clark (1985) made points which managers have had to discover in hard experience. For example, in an open access fishery, a total allowable catch (TAC) (at $F_{0.1}$ (see p. 263) or at any other management objective) by itself will still increase fishing capacity because fishermen compete with each other. Similarly, a simple regulation like a mesh size on its own may make fishermen become more efficient and hence increase fishing effort. Without a quota, limited entry can deplete a stock because competitors will fish against each other; for example, in the British Columbian salmon fishery the fishermen were licensed – but the purse seiners replaced the gill netters.

Clark (1985) felt compelled to write that efficiency was a laudable

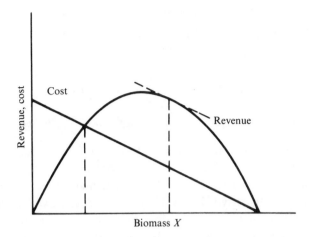

Figure 94. The simplest economic model as function of biomass; revenue is shown as a parabola and cost as an inverse line. The maximum economic yield is shown as the tangent to the parabola parallel to the cost curve, i.e. the greatest revenue above cost (Clark 1985).

objective. Most objectives are encompassed by a sole-owner model, which maximizes profit, if the word 'profit' is replaced by the phrase 'social welfare'. Graham (1943) said that he was expressing his law in socialist language when he wrote that unlimited fisheries became inefficient. In both systems, the manager must leave the fishermen free as far as he can. For this reason Clark (1985) wrote that to close down a depleted fishery 'can have unpleasant consequences for the fishing industry and consumers'. With this severe understatement, he noted that, at low catch levels, prices should respond to low supply; then catches should be reduced to ensure any degree of recovery. One of the troubles here is that most of the depleted fisheries are the pelagic ones for fish meal. If the collapse is due to recruitment overfishing, it has been difficult to avoid the advice to stop fishing, because the dependence of recruitment on parent stock is unknown.

Clark (1985) showed that quotas which are not allocated to men or vessels make fishermen compete too much. Further, he demonstrated that fishing effort can be stabilized by taxes or by transferable quotas, i.e. quotas which can be exchanged by sale. The economic formulation states that the price of the yield should be greater than the cost of the effort expended. So there are two fiscal measures which can be used, tax or licence. A tax increases the price of the yield and a licence increases the cost of the effort. The important point is that entry to the fishery is not limited by the EEZ. The only way to limit entry is by tax or by a licence by which quotas may be transferable.

The North American experience

The United States (Sissenwine and Marchessault 1985)

In 1945, President Truman proclaimed that the United States had the authority to establish conservation zones contiguous to their coastline (Rothschild 1983). In 1952, Chile, Peru and Ecuador laid claim to zones out to 200 miles. In early 1977, the United States, along with many other countries, extended its Economic Zone to 200 miles; in fact the zone was extended from three to two hundred miles under federal control and, up to three miles, stocks remained under state control (McHugh 1983). In 1976, the Fisheries Conservation and Management Act was passed and, under its aegis, eight Fishery Management Councils were established to prepare management plans for each major fishery.

The Act was based on multiple use and the management plans were to be formulated by officials, scientists, technical experts, representatives from industry and from the public. Plans were to be formulated in terms of optimum yield (or maximum sustainable yield (MSY), where necessary); defined as 'the amount of fish (a) which will provide the greatest overall benefit to the Nation, . . . [and] (b) will be prescribed on the basis of MSY as modified by any relevant economic, social or ecological factors'. Especial care had to be taken to conserve the biological basis of the stocks and at the same time to enhance the socio-economic viability of the fishing industry (Miles *et al.* 1982). The plans were discussed with maximum public and scientific input, including representation by recreational fisheries, conservationists, national economic protectionists, consumers and other marine interests (Smith 1982). Every special group was represented and the management objective must have become blurred.

The Pacific Management Council proposed optimal yields which were frequently adjusted for particular stocks. Gear restrictions were put forward 'without interfering with fishermen's freedoms' (Huppert 1985). Miles *et al.* (1982) gave tables of optimum yield, United States capacity, total allowable level for foreign fleets, and allocations to foreign fleets for various sectors. In the Bering Sea and Aleutian Is., the Americans caught 10000 tonnes of herring, but 1.33 million tons of other fishes were taken by Russians, Japanese, Koreans, Taiwanese and Poles in 1980. However, in the West Central Sector (between San Diego and Cape Flattery), the proportion of indigent catch increased from 21% in 1976 to 58% in 1983; such catches comprised three species of rock fish, some flatfish and roundfish (Francis 1985).

Perhaps the most important events took place in New England, governing the sea area between Long Is. Sound and the Canadian border. In the early 1970s, it was shown in ICNAF that the stocks of cod, haddock and yellowtail flounder were overexploited; indeed the George's Bank haddock may well have suffered from recruitment overfishing. In 1975 and 1976, new boats entered the fishery under high expectations from the Fisheries Conservation and Management Act, because they were to replace those foreigners who had departed (McHugh 1983).

The first groundfish management plan had three components: (*a*) minimum mesh sizes and minimum landing sizes, (*b*) spatial and seasonal closures to protect spawning haddock, (*c*) initial annual quotas of yellowtail flounder split down to voyage quotas. The later quotas for the haddock and yellowtail flounder were imposed to control by-catches in fisheries directed to other species. In 1977, 80% of the quotas of cod and yellowtail flounder were taken in the first half of the year. Directed fisheries for cod were closed and by-catch limits were imposed. In September of that year, cod by-catches for the rest of the year were estimated by vessel class. Further, the discarding of cod was prohibited so when the level of by-catch was reached the fishermen had to stop fishing or to discard fish illegally.

By 'Murphy's Law' the 1975 year classes of cod and haddock were large; indeed that of haddock was the biggest since 1963, as the stock recovered from recruitment overfishing, i.e. as fishing effort was reduced. Small fish were recruited to the adult fishery in the summer of 1977 and, of course, a large proportion might have been discarded. During the autumn of 1977 the quotas for 1978 were often modified and during that year there were enforcement failures; Dykstra and Wilson (1985) said that cod became 'pollack' and that remote ports were used. In March 1979, the catch quotas for cod and haddock were increased. The 1978 year class of haddock was also high. In August 1979 an Interim Plan was proposed by the Fishery Management Council, under which mesh regulations were improved and spawning areas were closed and in 1982 it was introduced. But management remained chaotic, with more closures and the voyage quotas being changed frequently.

By March 1982, there were no quotas, but minimum mesh sizes, minimal landing sizes and spawning area and season closures survived. In August 1983 the major object of management was to prevent stocks reaching levels that were too low. Fishermen had the freedom to switch from stock to stock, i.e. from quota to quota. By 1984, the haddock stock had declined again.

This short history of the New England management describes a failure. McHugh (1972) wrote that coastal state management might not be effective because sovereignty would be divided between state and federal government. Rothschild (1983) wrote:

constraining quotas generally increases falsification of fishing statistics and costly enforcement programs . . . There is a serious question as to whether the traditional fishery management paradigm – collect fishing statistics, set quotas, enforce quotas – can be cost effective under a regime where quotas actually constrain fishing and are strictly enforced . . . We now recognize that simply claiming authority over distant water fishing by coastal states was a piecemeal and simplistic solution to the problems of fishery management.

Open access remains possible in the waters managed by the coastal state.

However, the failure was an honourable one. The initial quotas were obviously low to obtain quick recovery of stock. The structure of enforcement was detailed and strict, so much so that evasion started more quickly than might have been expected. Managers were unlucky to encounter three strong year classes within such a short time. But entry was never limited because quotas were used without licences and no other fiscal measure was used.

Canada

North of the border, the Canadian Atlantic experience was quite different, because they were a little luckier. In the early 1960s, foreign fishing effort expanded sharply on the Canadian shelf. Templeman and Gulland (1965) wrote that there 'must be some direct control of the amount of fishing . . . [and] that presenting least difficulties is by means of catch quotas'. The quotas were introduced by ICNAF in 1970–2. In 1975, ICNAF was advised by its scientists to reduce the TACs for 1976, which they did to $F_{0.1}$ levels. At the same time, Canada announced that the 200-mile limit would be enforced before the 1976 meeting. But much more significantly, the Canadians asked for a 40% reduction in fishing effort by foreign fleets, as compared with 1973. Their justification was that the 'catch in 1973 was similar in quantity to that of the early 1960s, but that fishing effort had doubled and stock abundance declined by half' (May *et al.* 1979).

During the period of transition to coastal state management, the management objective, $F_{0.1}$, was used and catches and fishing effort were controlled by weekly returns. By 1980, there was evidence that the stocks had recovered to some degree. Licences had been used in the management of the Canadian fish fleet since 1973, but there had been a relatively heavy influx of boats of < 20 m length (May *et al.* 1979). However, MacDonald (1984) noted three problems: (*a*) dispute on how the national quota should be divided, (*b*) the international dispute with the United States on the 'median' line in the region of George's Bank, and (*c*) leakage of effort damaged the vessel licensing system (in other words, there was poaching and exchange of catches between licensed vessels). By 1982, there was considerable competition from fish imported from Iceland and Norway. In the Groundfish Plan of that year, a simulation model showed that the fishing effort was too high and an attempt was made to reduce it by allocating smaller quotas.

Some discontent with coastal state management led to the Canadian government setting up a Task Force on Atlantic Fisheries, chaired by Michael Kirby, in 1981. There were many recommendations of which the most important was that the licence was to be the property of the fishermen; it was to be transferable and licences were to be used to limit fishing effort. Further, mechanisms were to be created to foster greater contact between scientists and fishermen. Thus, two very important principles were brought to the fore, the use of an economic principle and the desire to win the consent of fishermen.

The Canadians were the first to use a multispecies model explicitly in management by quota. The cod stock in the Gulf of St Lawrence had declined as the mackerel stock increased. On the assumption that mackerel ate the cod larvae, Lett, Kohler and Fitzgerald (1975) modelled the changes in cod recruitment (in numbers) as a function of cod and mackerel biomass and larval abundance. The model was rather complex and was developed for the period 1954–72. Between 1975 and 1978, it was used in the Standing Committee on Research and Statistics, the scientific committee of ICNAF, but in 1979 it was replaced by a simple cohort analysis, perhaps because the estimated biomass at $F_{0.1}$ was higher. Doubleday and Beacham (1982) gave the statistical reasons for doubting the structure of the model.

The sharp reduction in effort in 1976 gave the Canadians a good start to the management of their coastal stocks. They had introduced a system of licensing into their groundfish fleets some years before. But within a few years the stocks were tending to decline, presumably because entry to the fishery remained fairly open. There is evidence now that in future entry will be more limited and that the consent of the fishermen will be sought.

The Scandinavian experience

Of the seven Scandinavian countries, four (Norway, Iceland, Greenland and the Faroe Is.) are coastal states *par excellence*. The fisheries are simple – one, two or three species at most; the foreign fishermen have nearly all departed and the administrations wish to obtain good conservation. The Norwegians share stocks in the Barents Sea with the Russians and in the North Sea with the European Economic Community (EEC). The Faroe Islanders share stocks with the Community and the Norwegians. The Icelanders do not share stocks with any other nation, although the recovering Atlanto-Scandian stock of herring will presumably enter their waters (other Scandinavian countries and the Belgians fish in Icelandic waters); similarly there are links between the Icelandic cod stock and that at west Greenland, but the latter is at present in low abundance. Hence the Icelanders have had the greatest chance of success in coastal state management.

Schopka (1980) described the changes since the EEZ was declared. In 1976, the mesh size was raised to 135 mm and the minimum landing size for cod was lifted to 50 cm (at the 50% retention length). In 1977, the mesh size was raised again to 155 mm and for haddock the minimum landing size was set at 45 cm. Such changes probably inflicted some loss to the fishermen in those years. Nursery grounds in the north-west and in the north and east of the island were closed to trawling. The result of these measures by 1980 was to reduce fishing mortality considerably between the ages of three to six. The maximum yield per recruit in 1976 was 1.65 g/R' (at $F = 0.7$) and in 1979 it had increased to 1.90 g/R' (at $F = 0.4$). This is the traditional cure for growth overfishing executed with no restraint from other nations or from conflicting fisheries. The same procedures were enforced in Norway and in the Faroe Islands with much the same success.

The north-east Atlantic

The development of structure (Leigh 1983)

Until 1977 the stocks of fish in the north-east Atlantic were managed by NEAFC. Today a large part is played by the European Commission. The Soviet Union, Eastern bloc and Scandinavian countries play complementary parts. The present role of the Commission is expressed in the Common Fisheries Policy, which was finally formulated in 1982. At an earlier stage, before the entry of Britain, Ireland and Denmark in 1972, the fisheries policy was simple: (*a*) each country had equal access to fishing grounds, (*b*) fleets should be restructured from Community funds, (*c*) there should be a market between countries, and (*d*) trade with third countries (outside the EEC) should be fostered.

Before the EEC was enlarged, its subsequent history was influenced by a referendum in Norway on whether to join the EEC and by the Cod Wars between Britain and Iceland. At this time the possibility of EEZs of 200 miles was being discussed in the long running Law of the Sea Conference. Norway did not join the Community in 1974 for two reasons: first, that the towns in the northern half of the country depend almost entirely upon fishing and, secondly, that the stocks of cod and haddock in the Barents Sea were exploited heavily by distant-water fleets from many countries. The second series of events had started when Iceland extended her limits from 3 to 4 miles in 1952 to 12 miles in 1958 and to 50 miles in 1971. In 1973 a two-year phasing out agreement was reached between Iceland and Britain and between Iceland and Belgium. In 1975, a 200-mile limit was declared because, in a statement to the United Nations, the Icelanders said that the cod stocks were depleted and that they had lost their exemption from duty in Community countries. During these events there was some naval activity. Icelandic expectations stemmed from the very high catches made in 1945 after the Second World War, when foreign trawlers had been absent from their shelf. Whereas the relaxation of fishing effort must have played a part, the stocks off Iceland had probably been enhanced considerably by a potent natural change which was not repeated (Cushing 1982). The consequences of the two events, the Norwegian refusal to join the EEC and the Cod Wars between Britain and Iceland, was that distant-water trawlers from Britain, France and the Federal Republic of Germany (amongst other countries) were gradually excluded from Icelandic waters and from the Barents Sea. The subsequent arguments on the Common Fisheries policy within the EEC really revolved around the share of catch for each country, after the distant-water fleets had been excluded from northern waters.

In February 1976, external relations in fisheries with countries outside the Community were transferred from member countries to the Community. The Commission was to propose annual TACs for each stock (on advice from ICES) and their allocation to member states and to third countries. Lastly, 140 million units of account were to be devoted to restructuring. Britain and Ireland rejected these proposals and countered with an exclusive limit of 12 miles and also a 50 mile limit in which they had predominant shares. In November 1976, the Hague compromise appeared: if the conservation regime were not established in 1977, member states could adopt appropriate measures to ensure the protection of resources in the fishing zones off their coasts. In effect the use of the Hague compromise was extended in later years, but the Commission had to be consulted at all stages of the procedure. Special regions were identified where the local communities were particularly dependent on fishing and their vital needs had to be taken into account in applying the Common Fisheries Policy (off Greenland, off Ireland and off the northern parts of Britain). Irish fisheries were to be expanded. On 1 January 1977, the EEC extended its limits to 200 miles and the USSR, Eastern bloc countries and Japan were excluded from the Community's EEZ. These events dominated the negotiations during the subsequent five or six years.

During this period Britain passed five national measures under the Hague compromise. Of these, by far the most important was the ban on the capture of herring in the waters around the British Isles. The capture of immature herring was banned and a minimum size of 21 cm in length was laid down. For the North Sea herring and those off the west coast of Scotland, the Commission approved this proposal. The stocks of herring in the North Sea had been heavily exploited since 1955, but in the northern North Sea the stocks had enjoyed an augmentation of recruitment (Burd 1978) during the 1950s and early 1960s. By 1970, all stocks had been depleted and in 1974 the Liaison Committee of ICES recommended a ban on capture which was not subsequently taken up by NEAFC. This recommendation was repeated by the Advisory Committee on Fish Stock Management (ACFM) in 1976. The ban was effective although not complete, and by 1982 stocks had recovered enough to allow a limited resumption of fishing. NEAFC had failed to conserve the herring, but the Community succeeded by approving the British national measure.

There were four other British national measures during the period: (a) an increase in mesh size of trawls for *Nephrops* in March 1977; (b) establishment of a licensing system for the Mourne and Isle of Man herring fisheries in July 1980; (c) the Norway pout box off the east coast of Scotland from which industrial fishermen were excluded in November 1977, with the Commission's approval (but an eastward extension was not approved); and (d) a further increase in mesh size for *Nephrops* in the Irish Sea in January 1979. On all four national measures the Commission proceeded against the United Kingdom in the European Court, judgement being entered against the United Kingdom. However, most of the science on which the national measures were based derived from ACFM recommendations.

The two measures on *Nephrops* were based on the fact that with larger meshes in their trawls the fishermen would obtain larger catches of bigger and more valuable animals after two or three years; but the extraordinary point was that, with smaller meshes, the fishermen discarded large quantities of small shrimps. The reason for the Norway pout box was that large quantities of small and immature haddock were caught in the Danish industrial fishery off the Scottish east coast. Indeed, in 1974, the tonnage of small haddock landed by industrial fisheries for feeding animals was greater than the tonnage landed for feeding people; in numbers, of course, the ratio was very much greater, by a factor of five or more. The British national measures constituted the major explicit steps taken in management between 1974 and 1982; the reasons for Court action lay not in the main thrust for conservation but in the details of presentation.

In addition to the British national measures, the main arguments in the development of the Common Fisheries Policy concerned access to British waters and the allocation of national quotas within the recommended TACs. One of the consequences of the extended limits was that the British claimed an element of sovereignty out to 200 miles, or median lines. Because 60% of EEC catches were made within the British zone, the British proposal for a predominant share for them within a 50 miles limit was the reason that their co-members of the Community denied it. The argument on access to British waters ended with some 6–12 mile sectors being retained, together with a box around Orkney and Shetland restricted for vessels larger than 24 m in length (which effectively restricted access to the box for vessels other than British).

The Law of the Sea Conference stated that the coastal state determined the TACs of stocks within its EEZ in order to prevent overexploitation and maintain stocks at the MSY. In EEC waters, the coastal state was the Community and in December 1976 the Scientific and Technical Committe was established by the European Commission to receive scientific advice from ICES (i.e. the annual report of ACFM). In the argument on allocation, the European Commission accepted jurisdictional losses (i.e. off Iceland and in the Barents Sea) and regional difficulties (for example off the Orkney Is. and the Shetland Is.). The final result of the argument on allocation was that Britain obtained 36% of catches, Denmark 24.5%, France 13%, Federal Republic of Germany 13%, Ireland 4.5%, Netherlands 7% and Belgium 2.5%.

In June 1982, the Commission established its rules for the coordination of national inspection by Community officials (of, for example, mesh sizes, minimum landing sizes, etc.) and introduced standard log books for all but the smaller vessels. Community inspectors were appointed. In

January 1983, rules were introduced for the control of national quotas as exhaustion of quotas is approached. During the decade 1973–82, 100 million écus were used on restructuring the Community fleets. In addition, the Community contributes half the costs of laying up vessels and scrapping, subsidizes exploratory voyages and encourages the development of inshore fisheries.

There is a considerable difference between the structure of NEAFC and that of the Community. In NEAFC, the scientific advice from ACFM was presented by its chairman to NEAFC itself. NEAFC comprised the national administrators for fisheries, each of which was supported by aides, scientists and representatives of industry. Agreements were reached around the table, limited by the political and industrial influences, but there was no means of enforcing them although there was common inspection. Conservation proceeded too slowly and stocks became depleted.

In the Community, scientific advice is received by the Scientific and Technical Committee from ACFM. However, it is important to remember that the Commission works with the Community through the Committee of Permanent Representatives (COREPER), which itself reports to the Council of Ministers. During the development of the Common Fisheries Policy, questions were and are frequently referred to the Council of Ministers. This structure is stronger than NEAFC for three reasons: first, Community regulations have the force of law and are directly applicable in all member states; secondly, there are Community inspectors; and thirdly the Commission can have resource to the European Court.

State of the stocks in the north-east Atlantic

Each year some twenty or more Working Groups of ICES report on the state of the stocks in the north-east Atlantic and their work is summarized in the ACFM report. Two management objectives were used, F_{max}, (F at MSY) and $F_{0.1}$. The TAC is the Community quota, subsequently divided nationally. In 1984, ACFM said that fishing mortality should be reduced to that at MSY in the Barents Sea cod and saithe stocks, the Irish Sea cod, plaice and sole stocks, North Sea haddock, cod, whiting, and saithe stocks, and the western mackerel stocks. The fishing mortality on North Sea sole should be reduced as quickly as possible. For North Sea plaice, 30% of the catches were not reported officially. The North Sea mackerel fishery should have been closed. Such were the results for some of the more important stocks in the north-east Atlantic.

Table 15. *Discards and industrial catches of cod, haddock and whiting in the North Sea*

Year	Total catch of cod, haddock and whiting in tons	Discards and industrial catches of cod, haddock and whiting in tonnes	Sum	Proportion of discards and industrial catches (%)
1974	426447	430555	857002	50.3
1975	410312	528926	939238	56.3
1976	449769	488478	938247	52.1
1977	392867	309225	702092	44.0
1978	435346	205930	641276	32.1
1979	430621	264130	694751	38.0
1980	442160	423426	865586	48.9
1981	520626	237642	758268	31.3
1982	494290	124290	618580	20.1

For the North Sea, some of these conclusions will be modified by the reports of the Multispecies Working Groups referred to in Chapter 14. Now that the natural mortality of juvenile fishes can be fairly well estimated, large changes in yield are not expected from possible changes in fishing effort of up to 40%; but this is really the consequence of the need for coexistence of fisheries to provide for human consumption and for industrial purposes. The management objective with a reduction of fishing effort in the North Sea is now the gain in profit set against the loss of jobs.

No account of the North Sea stocks is complete without a discussion of the Danish industrial fishery for fishmeal. The Danish fishermen started by exploiting immature herring after the Second World War. Later they turned to Norway pout, sprats and sandeels, species not exploited very much for human consumption. In 1974, catches from this fishery reached two million tonnes and quantities of small haddock and whiting were taken which might have grown to be caught in the human-consumption fishery. Another problem is the quantity of fish discarded in the trawl fishery because the distributions of young fish and adult fish overlap, particularly when a large year class appears as young fish. Table 15 shows the trends in quantities of cod, haddock and whiting caught in the industrial fishery or discarded.

in 1974 half the catches comprised fish for industrial purposes or discards. In 1982 the proportion was reduced to about one third or one fifth. The high proportion in 1980 was due to the apparent discard of 170675 tonnes of cod, which seems unlikely. The industrial catches of haddock and whiting have declined from 130 000–197 000 tonnes in 1974–6 to about 50 000 tonnes in 1982. The discards of these two species have declined from 270 000–390 000 tonnes in 1974–6 to about 70 000 tonnes in 1987. The decline in the proportion of discards and of industrial catches of cod, haddock and whiting are due to two causes: (*a*) the establishment of the Norway pout box off the east coast of Scotland; (*b*) enforcement of mesh size regulations.

The major success in the north-east Atlantic was the recovery of the herring stocks in 1982, after a ban since 1976. The total catch in 1983 amounted to 83 000 tonnes and the TAC for 1984 for the three stocks (northern, central and southern North Sea) amounted to 98 000 tonnes. The new year classes entering the North Sea herring stocks are large and there is every hope that a full recovery will take place. The markets, however, have not yet started to recover.

The system of management in the north-east Atlantic has not yet been successful because the Common Fisheries Policy took so long to develop. It remains true that the Scandinavian countries have made some progress towards good management. The assessment problem in the North Sea has been difficult for many years, but there is considerable hope now because the essential problems can be expressed in a new way. The new system of inspection should make it impossible to evade the report of catches, particularly when supported by the European Court.

The species commissions since 1977

There are three species commissions established under the inspiration of the United States, the International Halibut Commission, the International Pacific Salmon Commission, and the Inter-American Tropical Tuna Commission. The International North Pacific Commission was established primarily to negotiate the state of Pacific salmon stocks after the Abstention principle had been stated in 1953; in later decades the scope of this commission was broadened considerably. When the coastal state became the prime agent of management, some of the interests of the United States were taken from some of the species commissions. The functions of the Inter-American Tropical Tuna Commission remain as they were because in the Law of the Sea the tuna-like animals are 'highly migratory species', exempt from coastal management; it was shown long ago (Otsu 1960) that albacores cross the Pacific in a year or less. The International Commission for the Conservation of Atlantic Tunas has a similar function.

One of the consequences of the Fisheries Conservation and Management Act was that a new treaty was signed between the United States and Canada. Under this treaty the reciprocal agreement by which halibut fishermen could use either Canadian or United States waters was phased out in two years; the Halibut Commission was to continue. Its advice is channelled through the International North Pacific Fisheries Commission to the United States and to Canada. In recent years the science of the Halibut Commission has been sharply stimulated: Skud (1977) showed that the conclusion of Thompson and Herrington (1980) and Thompson and van Cleve (1936) that there were two halibut stocks was no longer true. Deriso and Quinn (1983) and Deriso (1985), with the use of sequential analysis and a delayed-difference production model (which takes into account the annual changes in recruitment), have allowed scientists to suggest reasonable quotas each year.

The 1953 convention, under which the International Pacific Fisheries Commission operates, was amended so that the commission should provide for scientific studies and a forum for cooperation in the collection of data; in other words, the commission became a regional scientific organization as required by the Law of the Sea. The Abstention principle remains. In 1981, the commission conducted a symposium on groundfish yields and Pacific cod biology: Leaman and Beamish (1984) showed that some redfish live as long as seventy years and Tanaka (1984) gives two methods of calculating catch quotas, including a replacement yield.

A major problem faces the Inter-American Tropical Tuna Commission. It has been known for a long time that there were distinct groups of yellowfin tuna along the equatorial system (Royce 1964). Suzuki *et al.* (1978) have shown that in fact there are three spawning groups, western, central and eastern, as shown by quarterly distributions of larvae. The fishery does not work far out on a Pacific scale so we should expect the stock to be defined quite properly in a short time. The International Commission for the Conservation of Atlantic Tuna has met biennially since 1971; minimum landing sizes have been agreed and catches of bluefin have been prohibited in the western Atlantic because of heavy fishing pressure. The Atlantic is a large ocean and the Commission has taken a long time to create an ocean-wide science but much good work has been done, particularly on stock identification.

Whaling and sealing

Immediately after the Second World War, part of the impetus to resume whaling lay in the world need for fat. Since the early 1950s most stocks, but perhaps not all, were overexploited and the markets no longer exist except in a few places. By the mid 1960s the Whaling Commission had found a way in which the stocks could be managed: the stock of blue whales (and at the same time the species) was protected.

Subsequently, many people turned against the killing of whales and their view was expressed both in the Scientific Committee and in the commission. On three occasions the management procedure was changed, finally to present a range of choice from the Scientific Committee to the commission. This considerable advance should have left the science to the Scientific Committee and the political choice to stop whaling to the commission. In a sense, this is what happened, for whaling has now stopped for a period, except in some places and for some people (for example the aboriginals who need to take 178 gray whales).

The coastal state

The drive for coastal state management arose from two sources. The first were those countries the continental shelves of which were exploited during the first industrialization, such as Iceland, Norway and Canada; but we recall that the Grand Banks were quite well exploited long before the first industrialization started (see Chapter 3). The second drive toward coastal state management arose from those countries predominantly equatorward of 40° latitude, the continental shelves of which were subject to the second exploitation.

Coastal states had managed fish stocks before the early months of 1977, prawns in the Gulf of Carpentaria, anchovy off Peru, and the prawn stocks off Western Australia. Limited entry had been practised in the latter fishery more or less since it started and in the Gulf of Carpentaria economic models have been used successfully in management. The collapse of the anchovy stock was a complex event but access was open in the waters of the coastal state.

The extension of limits to 200 miles raised hopes because it was thought to be enough to replace the limited actions of the international commissions with the potentially more decisive ones of the coastal state. In Scandinavian countries, progress was made towards good management. In Canadian waters vessels had been licensed since 1973, but the simple reduction in fishing effort was in the end not enough. Off New England the approach failed. In the waters of the EEC, the Common Fisheries Policy has not been working long but little has been achieved.

Fisheries that are unlimited become unprofitable and inefficient, wrote Graham (1943). This means that entry must be limited. In the prawn fisheries off Western Australia, licences were issued early in development. But when a stock is overexploited some fishermen should stop their dangerous trade. The initial delight of the Beverton and Holt (1957) solution to growth overfishing was that overfishing could be cured without putting fishermen out of work. Now we know that the pleasure was premature and must return to Graham's Law. Entry can be limited with licences and transferable quotas but the latter may be hard to administer. The objectives of management have become more complex, a compound of yield, value and jobs. Hence, the steps to be made should be short and made with the consent of the fishermen.

16

The provident sea

The sea has long provided food for men, but only in the last hundred years has the supply been threatened. In antiquity, the fish were probably taken in sight of the shore. Our image of preindustrial fisheries is perhaps ideal, distant in history, short of description and lit by old prints. Many were small and supplied coastal markets. Vessels were vulnerable and the many fishermen worked very hard. But the history of the herring and cod fisheries belies the simple image. The old herring fisheries, Scanian, East Anglian and Dutch, landed quite significant quantities which were used as a preserved food for all in winter and at all seasons by soldiers and travellers. The nineteenth-century Scottish fishery, with a preindustrial method of capture, was supported by the more general industrialization. The long history of the cod fishery, however, is that of the pre-industrial fishery, par excellence. It was a distant-water trans-atlantic fishery which carried dried and salted cod back to the catholic countries of Europe. In France, particularly, there were transport links from the major ports to the centres of population. Again, the fish was preserved for winter use and it provided a fair part of basic diet.

The industrial revolution started towards the end of the eighteenth century and continued in textiles, iron, steam, engineering, canals and railways well into the nineteenth century. In Europe and in North America, cities were expanded (or even built) to accommodate the factories and the people to man them. Demand was generated for food and railways were actually built to take fish to the cities, for example that to Grimsby in 1849. And the fisheries for the Pacific halibut did not really get under way until the trans-american railways were opened at the turn of the century.

Industrializaton had direct effects upon the fisheries before the mechanization of capture. Sardines were canned in France in the 1830s and nets were made by machine in Scotland at about the same time. Boats fished further away from port as the inshore stocks were thinned and so ice was needed and in the United Kingdom ice was made in Barking, Lowestoft and Grimsby; later in the Pacific north-west, freezing plants were established in Seattle and Vancouver to store the halibut. During the nineteenth century fishing vessels were gradually improved and became somewhat larger – Lowestoft smacks, Breton sardine and tunny boats. In the same period, in the waters north of Cape Hatteras, American fishermen slowly developed the purse seine in more or less the form used in the present century.

The industrial revolution stimulated fisheries in the North Sea and off the coasts of North America, as a result of the demand for fish in the burgeoning cities. But the most dramatic event in industrialization was the mechanization of capture. Steam had been used to power vessels in the early nineteenth century, but it did not drive fishing vessels until the last two decades of that century. Then came the steam purse seiner, the steam trawler, the steam drifter, the steam whale catcher and the steam longliner. Not all were equally effective, but the steam trawler was probably four times as efficient as the smack in catching fish. The steam whale catcher brought all the pelagic species of the great whales into capture with ultimately disastrous results. The steam drifter and the steam liner were more effective than their predecessors in that more nets or lines could be shot and the vessels could reach the markets more quickly. The purse seiner with the power block eventually became the most destructive engine for killing fish because the fishermen could not distinguish small fish from large before capture and a large mesh cannot be used. Thus, the industrial revolution in the fisheries became effective when the gears were mechanized and when the vessels were driven by steam and later by diesel oil. It was no accident that the steam trawler was equipped with the double-barrelled steam winch by which the sidewinder could shoot and haul the trawl in half an hour as compared with three hours by hand.

During the first industrialization, catch rates declined in the North Sea in the 1890s and those of the Pacific halibut were diminished by 1910. Yields in weight eventually suffered and many stocks suffered from growth overfishing, a problem which Petersen hoped would have been solved by the fishermen themselves searching for the larger fish. The problem of the Pacific halibut was solved in the 1930s in that yields and catch rates recovered as licences restrained access, if unfortunately with economic loss. The more general problem of growth overfishing in the Atlantic demersal stocks had to await a solution until the 1950s. In the last years of that decade Beverton persuaded the plaice fishermen of Lowestoft to work further afield for the larger fish.

The stocks of whales and seals were reduced from the time of first exploitation, in some stocks rather slowly and in others quite quickly. The stocks suffered from recruitment overfishing despite the fact that their potential natural rate of increase is of the order of 7% year^{-1}. Some stocks such as the North Atlantic right whale have been very nearly reduced to extinction, whereas others such as the Pacific gray whale have recovered in a healthy manner. It is not yet known why there is such a marked difference in the order of recovery. However, the fact remains that industrialization, which brought the great whales to capture, reduced all stocks to low levels, including those in the Antarctic.

The second industrialization started with the invention of the stern factory trawler and the power block purse seiner. The stern trawler opened a great market for frozen fish and at about the same time the markets for fish meal and oil appeared, served by trawlers and purse seiners. But the most important development was the distant-water exploration throughout the world led by the Russians and the Japanese. The latter put their main effort into the broad shelves of the North Pacific, and the Russians worked the continental shelves of the world ocean. The upwelling areas were exploited by many nations but the coastal state often took the major part, indeed the Peruvians took all the catches off their coast. In 1949, at the UNSCURR (see p. 258), Graham drew a chart of the world distribution of stocks, exploited and unexploited: most were not fished at all. Today he would have drawn a different chart, showing that nearly all stocks were exploited, a large proportion fully exploited or more so. Such was the result of the second industrialization which brought the world yield towards its greatest quantity.

One consequence of the second industrialization was that many pelagic stocks were sharply reduced by recruitment overfishing. The list of stocks is long and the losses to men and companies were great, primarily because the scientists did not understand the nature of the dependence of recruitment on parent stock. They still do not, but safeguards have been invented. So, both demersal and pelagic stocks throughout the world stand in need of the good management which we are only slowly learning.

However, the major consequence of the second industrialization was the extension of the Exclusive Economic Zones in the late 1970s. The reasons were twofold: (*a*) the appearance off the shores of Third World countries of distant-water vessels foreign to them, and (*b*) the hope that the coastal state had the power to control fisheries in a way that had evaded the international commissions. We have seen that the hope was not immediately realized because access in the coastal waters remained open. But the hope must be kept alive because only the coastal state has the power to take action.

In Britain the first institutional step during industrialization was the scrapping of the pre-industrial regulations which had not been enforced.

Indeed, Huxley's view was that the stocks in the sea were inexhaustible. However, the International Council for the Exploration of the Sea was founded in 1902 in the hope that the scientific examination of fish stocks would provide a language for the resolution of conflicts in fisheries between nations which had been common for centuries. The first treaty was that on the Pacific halibut between Canada and the United States in 1929 and it was based on Thompson's science and upon limited entry.

In the 1930s the scientific problems were well formulated by Russell, Thompson and Graham and their work culminated in the Overfishing Convention of 1946. Graham was disappointed that the Convention listed no commitment to a restraint on fishing effort, if it were needed. At about the same time, the International Whaling Commission was established in which the limitation of fishing effort was expressly forbidden.

The species commissions for halibut, salmon and tuna were established on the west coast of North America in the 1930s. In the North Atlantic, the more general commissions came into being in the 1950s. Beverton and Holt solved the problem of growth overfishing, Schaefer formulated the production model and the stock/recruitment problem was stated by Ricker and by Beverton and Holt. The two Atlantic Commissions established mesh sizes and minimum landing sizes and, later, quotas; the Atlantic Tuna Commission (ICCAT) was established in 1965. But the inevitable progress towards coastal state management was under way, even if tuna was excluded.

The major institutional steps were in the voluntary agreements to take common action. Today some of that action appears weak and late, but often at the time people thought that a major problem had been solved. When a solution is reached, however, a new problem arises: the least common mesh size in the North Atlantic preceded heavier exploitation and the MSY of yellowfin tuna was nearly doubled as the boats fished further away from port during the 1960s and 1970s. The coastal state had the power to take the action that the commissions could not reach, but sometimes that power was not enough. Each institutional step raised new problems and each ideal solution became submerged. If there is a rule, it is that the problems of management change with each step made. Graham's view was effectively a modern economic view, except that his successors would demand a sole-owner device, such as a tax or transferable licence. Such a quasi-economic objective has been well perceived since the time of Petersen. Graham taught us that there was a greatest yield, but, to him, to limit effort for greater value was much more important.

The management objectives of rational exploitation have changed in many ways. The maximum sustainable yield remains a desirable objective, if it is shown that gains in yield can still be made by a reduction in fishing effort. But the attainment of that objective must put fishermen out of work. When fisheries regulation was started generally in the North Atlantic, the problem of unemployment was avoided merely by the increase in mesh size. Today when mesh sizes in the North Atlantic can no longer be increased by very much, the rate at which an objective in yield or in value can be attained is governed by the degree of which the loss of jobs can be softened. The simple earlier objectives have been replaced by a more complex one, the greatest yield for the greatest value and the least loss of jobs.

When, in general, authority for regulation was transferred from international commissions to the coastal state, there was considerable hope that regulations would be enforced. However, access can remain open in coastal state water and the early hopes have not always been fulfilled. Off New England, management became chaotic and eventually the system of catch quotas had to be abandoned, perhaps because quotas without licences must lead to cheating. Off the Atlantic coast of Canada, hopes were higher because fishing effort had been much reduced in the year before the coastal state had become effective; a few years later some discontent was being expressed because fishing effort was probably too high, too many licences having been issued. The Scandinavians were luckier because their stock structures were simple and so the ages at first capture could be sharply increased. In the waters of the European Community there was little movement towards general control until 1982 when the Common Fisheries Policy was agreed. The most important event since then has been the establishment of a Community Inspectorate which will coordinate and oversee the national inspection systems. Such were the main trends of events since 1977; in other places there have been various attempts at control, for example that of the rock lobster in Western Australia.

The rock lobster fishery started in 1944. In the early 1960s an increment of effort produced no gain in catch and in 1963 the boats were licensed with a maximum number of pots per boat (Morgan 1980). Bowen and Chittleborough (1966) estimated a maximum sustainable yield. By the early 1970s, fishing effort was reduced by about a quarter. Bowen (1971) described a number of inefficiencies as consequences of limited entry. By 1980, a licence was worth about A$100 000. Despite the luck and administrative determination, catches and effort increased during the 1980s (Morgan 1980). It is the clearest example, since the time of Thompson on the Pacific halibut, of the principle that limited entry benefits the fishermen despite economic difficulties.

Progress towards good management is slow. In the 1940s managers did not believe that fishing effort should be restrained; the International Whaling Commission failed in its first two decades and the Overfishing Convention was disarmed. But the International Commissions succeeded in establishing control by mesh regulation (and minimum landing size) in the North Atlantic. But their voluntary nature did not prevent the later overexploitation which led to the introduction of quotas. It was hoped that the coastal state would have the power to make management work, but it was not at first successful because access could remain open in the waters of the coastal state.

The nature of management depends much on the quality of the science that supports it, Thompson on the halibut, Beverton and Holt on the trawl mesh sizes in the North Atlantic, or Gulland on the estimation of quotas. And, conversely, some of the failures in management during the second industrialization were due to the scientific failure to understand the dependence of recruitment on parent stock. Recently, in the field of multispecies assessment, a notable scientific advance was made in accounting properly for the natural mortality of the North Sea fishes. The recent Multispecies Working Group of the International Council for the Exploration of the Sea has shown that the gains in yield to be expected in most species with a 40% reduction in fishing effort in the North Sea are very low indeed. Similarly, little loss is to be expected with a 40% increase in effort. Hence the objective of the maximum sustainable yield has vanished in that limited instance.

In the more diverse multispecies fisheries of the tropical seas, the relationships are probably more complex than a simple accounting of predation. Then perhaps the only way to proceed is to apply the adaptive management proposed by Walters and his co-workers for the multifarious problems of the Pacific salmon fisheries. With the proviso that the consent of the fishermen remains paramount, such an experimental form of management might disentangle some of the difficult scientific problems. But this is of course an extreme view. Where the fisheries remain simple, as in Scandinavian waters, a reduction in fishing effort may still produce a gain in yield. Or to put it another way, if the Danish industrial fishery in the North Sea did not exist, gains in yield might still be made if fishing effort were reduced.

When mesh regulation was introduced in the North Sea in the late 1950s, the effect of the Danish industrial fishery was limited to the fishery for immature herring in the eastern North Sea. By 1974, two million tonnes of Norway pout, sandeel, sprat and herring were landed in addition to whiting and young haddock. It was these fisheries which provided much of the information on predation from the international investigation of gut contents. At the same time the industrial fisheries were included in the general structure of assessment.

There used to be a simple belief that, with the right collection of statistics, stocks could be managed directly with very little trouble. But in the last few decades as each step in management reveals new problems, the scientific advice changes and new steps in management are taken. This changing pattern of science and management is likely to persist because there are no pristine rules. But the manager's job remains the same, to obtain the greatest gain in value or in yield that she or he can, with the least loss of jobs.

To fish is to hunt, and poachers are the best of all hunters. Whatever the regulation, the fishermen will try to evade it to some degree and, in the course of this unsuccessful pursuit, they become more efficient. In any fishery to which entry is limited, the fishermen will continue to compete, as shown in the Western Australian rock lobster fishery. But to make a regulation work, the consent of the fishermen must be obtained.

In the last eighty years or so, the rational exploitation of the sea has been a persistent challenge. If the simple objectives of earlier decades have become more complex, the method of reaching them will probably remain that of Thompson and of Graham, to take small steps in management and observe the results. Such a method could work in the North Sea across the exchanges, small reductions in effort producing small but secure increases in value.

Down the centuries fishermen have been lost at sea with their vessels. Today the ships are bigger, the radio is protective and the wind can be foreseen, but still from time to time, vessels fail to return from sea. The sea remains a dangerous place to work. The value that accrues to the fishermen is too low and the only way to preserve their lives is to pursue the rational exploitation of the sea.

REFERENCES

Aasen, O., Andersen, K. P., Gulland, J., Madsen, K. P. and Sahrhage, D. (1961). I.C.E.S. herring tagging experiment in 1957 and 1958. *Rapp. Procès-Verb. Cons. Int. Explor. Mer.* **152**.

Agger, P., Boetius, I. and Lassen, H. (1973). Error in the virtual population analysis; the effect of uncertainties in the natural mortality coefficient. *J. Cons. Int. Explor. Mer.* **35**: 93.

Aguilar, A. (1981). The Black Right whale *Eubalaena glacialis* in the Cantabrian Sea. *Rep. Int. Whal. Commn*, **31**: 457–9.

Allen, K. R. (1966). Some methods for estimating exploited populations. *J. Fish. Res. Bd Can.* **23**: 1553–70.

Allen, K. R. (1979). Towards an improved whale management procedure. *Rep. Int. Whal. Commn*, **29**: 143–6.

Allen, K. R. (1980). *Conservation and management of whales*. Univ. Washington Press, Seattle.

Allen, K. R. and Kirkwood, G. P. (1978). Simulation of Southern hemisphere sei whale stocks. *Rep. Int.Whal. Commn*, **28**: 151–8.

Allen, R. L. (1975). A life table for harp seals in the North West Atlantic. *Rapp. Procès-Verb. Cons. Int. Explor. Mer*, **169**: 303–11.

Alward, G. L. (1907). *Scientific investigation of the North Sea*. Grimsby News, Grimsby.

Alward, G. L. (1911). *The development of the British Fisheries during the nineteenth century with special reference to the North Sea*. Grimsby News, Grimsby.

Alward, G. L. (1932). *The sea fisheries of Great Britain and Ireland*. Albert Gait, Grimsby.

Andersen, K. P. and Ursin, E. (1977). A multispecies extension of the Beverton and Holt theory of fishing with accounts of phosphorus circulation and primary production. *Meddr. Danm. Fisk. og Havunders*, **7**: 319–435.

Anderson, E. D. and Paciorkowski, A. J. (1980). A review of the North West Atlantic mackerel fishery. *Rapp. Procès-Verb. Cons. Int. Explor. Mer*, **177**: 175–211.

Anderson, J. (1785). *An account of the present state of the Hebrides and western coasts of Scotland*. Elliot, Edinburgh.

Andersson, K. A. and Molander, A. R. (1928). The plaice fishery and the plaice in the Arcona basin and the Bornholm area (southern Baltic) during the period 1921–7. *Rapp. Procès-Verb. Cons. Int. Explor. Mer*, **48**: 19–30.

Anon. (undated). *A book for the sea side*. Religious Tract Society, London.

Anon. (1866). *Report of the Commissioners appointed to inquire into the sea fisheries of the United Kingdom*. HMSO, London.

Anon. (1893). *Report of the Select Committee on Sea Fisheries*, HMSO, London.

Anon. (1902). Meeting of Committee B. *Rapp. Procès-Verb. Cons. Int. Explor. Mer*, 1: 97–110.

Anon. (1904). Meeting of Committee B. *Rapp. Procès-Verb. Cons. Int. Explor. Mer*, 2: 6–44.

Anon. (1907). *Rapp. Procès-Verb. Cons. Int. Explor. Mer*, 7: 54–152.

Anon. (1920). *Report of the Interdepartmental Committee on the Research and Development in the dependencies of the Falkland Islands*. HMSO, London.

Anon. (1921). The history of trawling. *Fish Trades Gazette*, London, 19 March: 21–71.

Anon. (1953). *Third Annual Meeting International Commission on North West Atlantic Fisheries*, Dartmouth, Nova Scotia.

Anon. (1957). International Fisheries Convention 1946; Report of the Ad Hoc Committee established at the Fourth Meeting of the Permanent Commission, September 1955. *J. Cons. Int. Explor. Mer*, 23: 7–37.

Anon. (1959). Report of the Scientific Committee. *Rep. Int. Whal. Commn*, 10.

Anon. (1960). International Fisheries Convention of 1946; Committee on Mesh Difficulties. *Rapp. Procès-Verb. Cons. Int. Explor. Mer*, 151.

Anon. (1965). Environmental Symposium. *Int. Commn North West Atl. Fish Spec. Publ.* 6.

Anon. (1980) *Report on investigations during 1973–76*. North Pac. Fur Seal Commn, Washington, DC.

Anon. (1984). *Proc. 27 Annual Meeting North Pacific Fur Seal Commission, April 1984*, Washington, DC.

Anon. (1984). *Rep. Int. Whal. Commn*, 34.

Anon. (1986a). Report of the Ad Hoc Multispecies Working Group. ICES CM Assess. 9. Mimeo.

Anon. (1986b). Report of the Working Group on methods of fish stock assessment. ICES CM 1986 Assess. 10. Mimeo.

Anthony, V. C. and Waring, G. (1980). The assessment and management of the George's Bank herring fishery. *Rapp. Procès-Verb. Cons. Int. Explor. Mer*, 177: 72–111.

Baker, R. C., Wilke, F. and Baltzo, C. H. (1970). *The northern fur seal*. Circ. 336 Dept. Commerce US Fish and Wildlife Service.

Bakun, A. and Parrish, R. (1980). Environmental inputs to fishery population models for eastern boundary currents. *Int. Ocean. Commn Workshop Rep.*, 28: 67–104.

Baranov, F. I. (1918). On the question of the biological basis of fisheries. *Nauch. issledov. ikhtiol. Inst. Isv.* 1: 81–128.

Barrow, J. (1906). *Captain Cook's Voyage of Discovery*. Dent, London.

Bean, T. H. (1887). The cod fishery of Alaska. In *The fisheries and fishing industry of the United States*, sect. V *The history and methods of the fisheries*, ed. G. Brown Goode, vol. 1, pp. 327–415. Govt Printing Offices, Washington, DC.

Beaujon, A. (1884). The History of Dutch sea fisheries; their progress and revival especially in connection with the legislation on fisheries in earlier and later times. *Int. Fish. Exhib. Prize Essays*, II.

Beddington, J. R. and Cooke, J. G. (1981). Development of an assessment technique for male sperm whales based on the use of length data from the catches with special reference to the North West Pacific stock. *Rep. Int. Whal. Commn*, **31**: 747–60.

Beddington, J. R. and Williams, H. A. (1980). *The status and management of the harp seal in the North West Atlantic: a review and evaluations*. US Marine Mammal Commn, Washington, DC.

Benham, H. (1979). *The cod bangers*. Essex County Newspapers, Colchester.

Best, P. B. (1983). Sperm whale stock assessments and the relevance of historical whaling records. *Rep. Int. Whal. Commn Spec. Rep.* **5**, 41–56.

Beverton, R. J. H. (1962). Long term dynamics of certain North Sea fish populations. In *The exploitation of natural animal populations*, ed. E. D. Le Cren and M. W. Holdgate, pp. 242–64. Blackwell, London.

Beverton, R. J. H. and Holt, S. J. (1957). On the dynamics of exploited fish populations. *Fish. Invest. Lond.* ser. 2, **19**.

Beverton, R. J. H. and Hodder, V. M. (eds.) (1962). Report of working group of scientists on fishery assessment in relation to assessment problems. *Ann. Proc. Int. Commn North West Atl. Fish. Suppl.* **2**.

Blegvad, H. (1928). On the influence of the fishery on the stock of plaice in the Baltic proper. *Rapp. Procès Verb. Cons. Int. Explor. Mer*, **48**: 27–49.

Bockstoce, J. R. (1980). A preliminary estimate of the reduction of the western arctic bowhead whale population by the pelagic whaling industry, 1848–1915. *Mar. Fish. Rev.* Sept.–Oct. 1980, **42**: 20–1.

Bonner, W. N. (1982). *Seals and man*. Washington Sea Grant Programme, Seattle.

Bostrøm, O. (1955). 'Pedder Ronnestad' Ekkolodding-og meldetjeneste av Skreiforekomstene i Lofoten i tiden, 1 March–2 April 1955. *Prakt. fiskeforsok 1954 og 1955 Arsberet. vdekomm. Norges. Fisk.* **9**: 66–70.

Bowen, B. K. (1971). Management of the Western Rock Lobster (*Panulirus longipes cygnus* George). *Proc. Indo-Pac. Fish Council 14th Session*, **3**: 139–53.

Bowen, B. K. and Chittleborough, R. G. (1966). Preliminary assessment of stocks of Western Australian crayfish, *Panulirus cygnus* George. *Austr. J. Mar. Fresh. Res.* **17** (1): 93–122.

Bowen, W. D. and Sergeant, D. E. (1982). Mark–recapture estimates of harp seal production in the North West Atlantic. *Can. J. Fish. Aq. Sci.* **40**: 728–42.

Brander, K. M. (1977). *Management of Irish Sea fisheries – a review*. Min. Agric. Fish. Food Lab. Leaflet no. 36, Lowestoft.

Brander, K. (1985). How well do working groups predict catches? ICES CM 1985 G50. Mimeo.

Breiwick, J. M. and Mitchell, E. D. (1983). Estimated initial population size of the Bering Sea stock of bowhead whales (*Balaena mysticetus*) from logbook and other catch data. *Rep. Int. Whal. Commn Spec. Issue*, **5**: 147–51.

Breiwick, J. M., Mitchell, E. D. and Chapman D. G. (1981). Estimated initial population of the Bering Sea stock of bowhead whale, *Balaena mysticetus*: an iterative method. *Fish. Bull. U.S. Dept of Commerce*, **78**: 843–54.

Broadhead, G. C. (1959). Morphometric comparisons among yellowfin tuna, *Neothunnus macropterus*, from the eastern tropical Pacific Ocean. *Bull. Inter-Amer. Trop. Tuna Commn*, **3**: 355–91.

Broadhead, G. C. and Barrett, I. (1964). Some factors affecting the distribution and apparent abundance of yellowfin and skipjack tuna in the eastern Pacific Ocean. *Bull. Inter-Amer. Trop. Tuna Commn*, **81**: 419–73.

Brothers, E. S., Mathews, C. P. and Lasker, R. (1976). Daily growth increments in otoliths from larval and adult fishes. *Fish. Bull. Nat. Mar. Fish. Serv. US*, **74**: 1–8.

Brown, B. E., Brenhan, J. A., Grosslein, M. D., Heyerdahl, E. G. and Hennemuth, R. G. (1975). The effect of fishing on the marine finfish biomass in the North West Atlantic from the Gulf of Maine to Cape Hatteras. *Res. Bull. Int. Commn North West Atl. Fish.* **12**: 49–68.

Brown, Goode, G. (1879). The natural and economic history of the American menhaden. *U.S. Commn. Fish and Fisheries Rep. Commiss. for 1877*: pp. 1–56.

Brown Goode, G. (ed.) (1887*a*). *The fisheries and fishing industry of the United States*, sect. V *The history and methods of the fisheries*. Govt Printing Office, Washington, DC.

Brown Goode, G. (1887*b*). The swordfish industry. In *The fisheries and fishing industry of the United States*, sect. V *The history and methods of the fisheries*, ed. G. Brown Goode, vol. 1, pp. 315–36. Govt Printing Office, Washington, DC.

Brown Goode, G. and Clark, A. H. (1887). The menhaden fishery. In *The fisheries and fishing industry of the United States*, sect. V *The history and methods of the fisheries*, ed. G. Brown Goode, vol. 1, pp. 327–415. Govt Printing Offices, Washington, DC.

Brown Goode, G. and Collins, J. W. (1887*a*). The fresh halibut fishery. In *The fisheries and fishing industry of the United States*, sect. V *The history and methods of the fisheries*, ed. G. Brown Goode, vol. 1, pp. 3–89. Govt Printing Offices, Washington, DC.

Brown Goode, G. and Collins, J. W. (1887*b*). The Bank handline cod fishery. In *The fisheries and fishing industry of the United States*, sect. V *The history and methods of the fisheries*, ed. G. Brown Goode, vol. 1, pp. 123–33. Govt Printing Offices, Washington, DC.

Brown Goode, G. and Collins, J. W. (1887*c*). The Labrador and Gulf of St Lawrence cod fisheries. In *The fisheries and fishing industry of the United States*, sect. V *The history and methods of the fisheries*, ed. G. Brown Goode, vol. 1, pp. 133–47. Govt Printing Offices, Washington, DC.

Brown Goode, G. and Collins, J. W. (1887*d*). The Bank trawl line cod fisheries. In *The fisheries and fishing industry of the United States*, sect. V *The history and methods of the fisheries*, ed. G. Brown Goode, vol. 1, pp. 148–87. Govt Printing Offices, Washington, DC.

Brown Goode, G. and Collins, J. W. (1887*e*) The George's Bank cod fisheries. In *The fisheries and fishing industry of the United States*, sect. V *The history and methods of the fisheries*, ed. G. Brown Goode, vol. 1, pp. 187–98. Govt Printing Offices, Washington, DC.

Brown Goode, G. and Collins, J. W. (1887*f*). The haddock fishery of New England. In *The fisheries and fishing industry of the United States*, sect. V *The history and methods of the fisheries*, ed. G. Brown Goode, vol. 1, pp. 234–41. Govt Printing Offices, Washington, DC.

Brown Goode, G. and Collins, J. W. (1887*g*). The mackerel fishery. In *The fisheries and fishing industry of the United States*, sect. V *The history and* *methods of the fisheries*, ed. G. Brown Goode, vol. 1, pp. 247–313. Govt Printing Offices, Washington, DC.

Brown Goode, G., Collins, J. W., Earll, R. E. and Clark, A. H. (1884). Materials for a history of the mackerel fishery. *Rep. Commn Fish and Fisheries* IV: 93–531.

Buckland, F., Walpole, S. and Young, A. (1878). *Report presented to parliament on the herring fisheries of Scotland*. HMSO, London.

Bückmann, A. (1932). Die Frage nach der Zweckmässigkeit des Schutzes die Voraussetzungen für ihre Beantwortung. *Rapp. Procès-Verb. Cons. Explor. Mer*, **80**.

Burd, A. C. (1978). Long term changes in North Sea herring stocks. *Rapp. Procès-Verb. Cons. Int. Explor. Mer*, **172**: 137–53.

Burkenroad, M. (1948). Fluctuations in abundance of Pacific halibut. *Bull. Bingh. Ocean. Coll.* **11**.

Butterworth, D. S. (1983). Assessment and management of pelagic stocks in the southern Benguela region. Proc. Expert Consult. Costa Rica, ed. G. D. Sharp and J. Csirke, *FAO Fish. Rep.* **29**, vol. 1: 329–406.

Cadoret, B., Duviard, D., Guillet, J. and Kerisit, M. (1978). *Ar Vag, voiles au travail en Bretagne Atlantique*. Editions des quatre seigneurs, Grenoble.

Capstick, C. K., Lavigne, D. M. and Ronald, K. (1976). *Population forecasts for the North West Atlantic harp seals,* Pagophilus groenlandicus. ICNAF res. doc. 76/X/132.

Chambers, J. (1829). *A general history of the county of Norfolk intended to convey all the information for a general tour*, 2 vols. Longman, Rees, Orme and Browne, London.

Chapman, D. G. (1961). Population dynamics of the Alaska Fur Seal herd. *26th North American Wildlife Conf.* pp. 356–69.

Chapman, D. G. (1964). A critical study of Pribilov fur seal population estimates. *Fish. Bull.* **63**: 657–69.

Chapman, D. G. (1973). Management of international whaling and north Pacific fur seals: implications for fisheries management. *J. Fish. Res. Bd Can.* **30**: 2419–26.

Chapman, D. G. (1981). Evaluation of marine mammal population models. in *Dynamics of large mammal populations*, ed. C. W. Fowler and T. D. Smith, pp. 277–96. John Wiley and Sons, New York.

Chapman, D. G. (1983). Some considerations on the status of stocks of southern minke whales. *Rep. Int. Whal. Commn*, **33**: 311–14.

Chapman, D. G., Allen, K. R. and Holt, S. J. (1964). Report of the Committee of Three Scientists on the special scientific investigation of the Antarctic whale stocks. *Rep. Int. Whal. Commn* **14**: 32–106.

Chapman, D. G., Allen, K. R., Holt, S. J. and Gulland, J. A. (1965). Report of the Committee of Four Scientists on the special scientific investigation of the Antarctic whale stock. *Rep. Int. Whal. Commn* **15**: 47–63.

Chapman, D. G., Myhre, R. J. and Southward, G. M. (1962). Utilization of Pacific halibut stocks; estimation of maximum sustainable yield. *Rep. Int. Halibut Commn*, **31**.

Clark, A. H. (1887a). The whale fishery. I. History and present conditions of the fishery. In *The fisheries and fishing industry of the United States*, sect. V *The his-*

tory and methods of the fisheries, ed. G. Brown Goode, vol. 2, pp. 3–293. Govt Printing Offices, Washington, DC.

Clark, A. H. (1887b). The North Atlantic seal fishery. In *The fisheries and fishing industry of the United States*, sect. V *The history and methods of the fisheries*, ed. G. Brown Goode, vol. 2, pp. 474–83. Govt Printing Offices, Washington, DC.

Clark, A. H. (1887c). The fisheries of Massachusetts. In *The fisheries and fishing industry of the United States*, sec. II *A geographical review of the fishing industries and fishing communities in the year 1880*, vol. 2, pp. 113–280. Govt Printing Offices, Washington, DC.

Clark, C. W. (1973). The economics of overexploitation. *Science*, **181**: 630–4.

Clark, C. W. (1985). *Bioeconomic modeling and fisheries management*. Wiley Interscience, New York.

Clark, C. W. and Kirkwood, G. P. (1979). Bioeconomic model of the Carpentaria prawn fishery. *J. Fish. Res. Bd Can.* **36**: 1304–12.

Clark, F. N. and Marr, J. C. (1955). Population dynamics of the Pacific sardine. *Progr. Rep. Cal. Coop. Oceanic Fish. Invest. 1953–5*: pp. 11–48.

Collins, J. W. (1887). The shore fisheries of the southern Delaware. In *The fisheries and fishing industry of the United States*, sect. I *Natural history of aquatic animals*, ed. G. Brown Goode, vol. 1, pp. 527–41. Govt Printing Offices, Washington, DC.

Collins, J. W. and Rathbun, R. (1887). The seafishing grounds of the eastern coast of North America. In *The fisheries and fishing industries of the United States*, sect. III *The fishing grounds of North America*, ed. G. Brown Goode, pp. 5–78. Govt Printing Offices, Washington, DC.

Cooke, J. G. and Beddington, J. R. (1982). Western North Pacific estimates using the revised Beddington and Cooke technique. *Rep. Int. Whal. Commn*, **32**: 143–5.

Cooke, J. G. and de la Mare, W. K. (1983). An analysis in the trends in catch per effort for the North Pacific sperm whale with reference to the length structure of the catches. *Rep. Int. Whal. Commn*, **33**: 269–74.

Coull, J. R. (1983). The herring fishery in Shetland in the first half of the nineteenth century. *Northern Scot.* **5**: 123–40.

Crawford, R. J. M., Shelton, P. A. and Hutchings, L. (1980). Implications of the availability, distribution and movements of pilchard (*Sardinops ocellata*) and anchovy (*Engraulis capensis*) for assessment and management of the South African purse seine fishery. *Rapp. Procès-Verb. Cons. Int. Explor. Mar.* **177**: 355–73.

Crutchfield, J. (1965). *The fisheries: problems in resource management*. Univ. of Washington Press, Seattle.

Cushing, D. H. (1968). The East Anglian herring fishery in the eighteenth century. *J. Cons. Int. Explor. Mer.* **31**: 323–9.

Cushing, D. H. (1970–1). Kipling and the White Seal. *The Arlington Quarterly*, **3**: 171–88.

Cushing, D. H. (1973). Dependence of recruitment on parent stock. *J. Fish. Res. Bd Can.* **30**: 1965–76.

Cushing, D. H. (1975). *Science and the Fisheries*, Edward Arnold, London.

Cushing, D. H. (1982). *Climate and fisheries*, Academic Press, London.

Cushing, D. H. (1984). The gadoid outburst in the North Sea. *J. Const. Int. Explor. Mer.* **41**: 159–66.

Cushing, D. H. and Bridger, J. P. (1966). The stock of herring in the North Sea and changes due to fishing. *Fish. Invest. Lond.* ser. 2, **25**.

Cushing, D. H. and Harris, J. G. K. (1973). Stock and recruitment and the problem of density dependence. *Rapp. Procès-Verb. Cons. Int. Explor. Mer.* **164**: 142–55.

Cutting, C. L. (1955). *Fish Saving*. Leonard Hill, London.

Daan, N. (1973). A quantitative analysis of the food intake of North Sea cod, *Gadus morhua*. *Neth. J. Sea. Res.* **6**: 479–517.

Daan, N. (1978). Consumption and production in North Sea cod (*Gadus morhua*): an assessment of the ecological status of the stock. *Neth. J. Sea. Res.* **9**: 24–55.

Dade, E. (1933). *Sail and oar*. Dent, London.

Dahl, K. (1907). The scales of the herring as a means of determining age, growth and migration. *Rep. Norweg. Fish. Mar. Invest.* **2**: 1–36.

Davis, F. M. (1936). Mesh experiments with trawls, 1928–1933. *Fish. Invest. Lond.* ser. 2, **14**.

de Caux, J. W. (1881). *The herring and the herring fishery*. Hamilton Adams, London.

de Jong, C. (1972). Geschiedenis van de oude Nederlandse Walvisvaart. In *Deel een, grundslagen, onstaan en Opkomst 1612–1642*, pp. 123–50. Univ. of South Africa, Pretoria.

de Jong, C. (1978). *A short history of old Dutch whaling*. Univ. of South Africa, Pretoria.

de Jong, C. (1983). The hunt of the Greenland whale: a short history and statistical sources. *Rep. Int. Whal. Commn Spec. Issue* **5**: 83–106.

de la Morandière, Ch. (1962–6). *Histoire de la pêche française de la Morue dans l'Amérique septentrionale*, 3 vols. Maisonneuve and Larose, Paris.

Defoe, D. (1724–6). *Tour through the whole Island of Great Britain*, Peter Davies, Covent Garden, London.

DeLury, D. B. (1947). On the estimation of biological populations. *Biometrics*, **3**: 145–67.

Deriso, R. B. (1980). Harvesting strategies and parameter estimation for an age structured model. *Can. J. Aq. Sci.* **37**: 268–82.

Deriso, R. B. (1985). Stock assessment and new evidence of density dependence. In *Fisheries dynamics; harvest, management and sampling*, ed. P. Mundy, T. S. Quinn and R. B. Deriso, pp. 49–50. Univ. of Washington Press, Seattle.

Deriso, R. B. and Quinn, T. J. (1983). The Pacific halibut resource and fishery in regulatory area 2. Estimates of biomass, surplus production and reproductive value. *Int. Pac. Halibut Commn Sci. Rep.* **67**: 55–89.

Devold, F. (1963). The life history of the Atlanto-Scandian herring. *Rapp. Procès-Verb. Cons. Int. Explor. Mer.* **154**: 98–108.

Doubleday, W. G. (1979). A sensitivity analysis of selected catch projections. ICES CM 1979: G53. Mimeo.

Doubleday, W. G. and Beacham, T. D. (1982). Southern Gulf of St Lawrence: a review of multispecies models and management advice. *Can. Spec. Publ. Fish. Aq. Sci.* **59**: 133–40.

Doubleday, W. G. and Bowen, W. D. (1980). *Inconsistencies in reading the age of harp seals* (Pagophilus groenlandicus) *teeth, their consequences and a means of reducing resulting biases*. NAFO S.C.R. Doc 80/ 11/ 160.

Dudley, P. (1724). An essay on the natural history of whales. *Phil. Trans. Roy. Soc.* **33** (387): 256–69.

Duhamel du Monceau (1769). *Traité général des Pesches et histoire des Poissons*, vols. 1–3. Saillant et Nyon, Desaint, Paris.

Dunlop, J. (1978). *The British Fisheries Society, 1786–1893*. John Donald, Edinburgh.

Dykstra, J. J. and Wilson, J. A. (1985). East Coast groundfish experience: industry perspective. In *Fisheries management: issues and options*, pp. 223–33. Alaska Sea Grant Rep. no. 85.

Dyson, J. (1977). *Business in Great Waters*. Angus and Robertson, London and Sydney.

Earll, R. E. (1887*a*). The herring and the sardine fishery. In *The fisheries and fishing industry of the United States*, sect. I *Natural history of aquatic animals*, ed. G. Brown Goode, vol. 1, pp. 527–41. Govt Printing Offices, Washington, DC.

Earll, R. E. (1887*b*). The Spanish mackerel fishery. In *The fisheries and fishing industry of the United States*, sect. I *Natural history of aquatic animals*, ed. G. Brown Goode, vol. 1, pp. 543–57. Govt Printing Offices, Washington, DC.

Earll, R. E. (1887*c*). The mullet fishery. In *The fisheries and fishing industry of the United States*, sect. I *Natural history of aquatic animals*, ed. G. Brown Goode, vol. 1, pp. 553–82. Govt Printing Offices, Washington, DC.

Eberhardt, L. L. (1981). Population dynamics of the Pribilov fur seals. In *Dynamics of large mammal populations*, ed. C. W. Fowler and T. D. Smith, pp. 197–220. John Wiley and Sons, New York.

Elliott, H. W. (1884). The habits of the fur seal. In *The fisheries and fishing industry of the United States*, sect. I *Natural history of aquatic animals*, ed. G. Brown Goode, vol. 2, pp. 75–118. Govt Printing Offices, Washington, DC.

Elliott, J. M. (1985). The choice of a stock-recruitment model for migratory trout, *Salmo trutta*, in an English Lake District stream. *Arch. Hydrobiol.* **104**: 145–68.

Fink, B. D. and Bayliff, W. H. (1970). Migration of yellowfin and skipjack tuna in the eastern Pacific Ocean as determined by tagging experiments, 1952–1964. *Bull. Inter-Amer. Trop. Tuna Commn*, **8**: 15–109.

Fiscus, C. H., Baines, G. A. and Wilke, F. (1964). Pelagic fur seal investigations, Alaska waters, 1962. US Dept Commerce *Spec. Sci. Rep. Fish Wildl. Serv.* **475**.

Fisher, R. A. (1925). *Statistical methods for research workers*. Oliver and Boyd, London.

Foerster, R. E. (1936). The return from the sea of sockeye (*Oncorhynchus nerka*) with special reference to percentage survival, sex proportions and progress of migration. *J. Biol. Bd Can.* **3**: 26–42.

Forbes, E. (1841). *A history of British starfishes*. Jan von Voorst, London.

Forbes, E. (1849). *The natural history of European Seas*. Jan van Voorst, London.

Francis, R. C. (1985). Fisheries research and its application to west coast groundfish management. In *Fisheries management: issues and options*, pp. 285–336. Alaska Sea Grant Rep. 85.

Fry, F. E. J. (1949). Statistics of a Lake Trout fishery. *Biometrics*, **5**: 27–67.

Fukada, Y. (1962). On the stocks of halibut and their fisheries in the north east Pacific. *Int. North Pac. Fish Commn Bull.* **7**: 39–50.

Fukuhara, F. M., Murai, S., Lalanne, J.-J. and Sribhibhadh, A. (1962). Continental origin of red salmon as determined from morphological characters. *Bull. Int. North Pac. Fish. Commn*, **8**: 15–109.

Fulton, T. W. (1892). On overfishing and the culture of sea fish. *10th Ann. Rep. Fish Bd Scot.*: 171–93.

Fulton, T. W. (1908). A review of the fishery statistics for Scotland. *27th Ann. Rep. Fish Bd Scot.*: 129–91.

Fulton, T. W. (1911). *The sovereignty of the sea*. Blackwood, Edinburgh.

Garrod, D. J. (1977). The North Atlantic cod. In *Fish Population Dynamics*, ed. J. A. Gulland, pp. 216–42. John Wiley, London.

Garstang, W. (1900). The impoverishment of the sea. *J. Mar. Biol. Assn UK* N.S. **6**: 1–69.

Garstang, W. (1909). The distribution of the plaice in the North Sea, Skagerak, and Kattegat, according to age size and frequency. *Rapp. Procès-Verb. Cons. Int. Explor. Mer.* **11**: 65–134.

Gentleman, T. (1808). *England's way to win wealth and to employ ships and mariners*. London.

Gillingwater, E. (1790). *An historical account of the ancient town of Lowestoft*. Robinson and Nichols, London.

Goodey, C. (1976). *The first hundred years: the story of Richards' Shipbuilders*, Boydell, Ipswich.

Goodman, D. (1982). Analysis of harp seal management models. Center Env. Educ., Washington, DC.

Gordon, H. S. (1954). The economic theory of a common property resource. *J. Polit. Econ.* **62**: 124–42.

Gordon Clark, J. (1981). Objectives for the management and conservation of marine mammals. *FAO Fish Ser.* 5 (111): 103–16.

Gosse, P. H. (1846). *The oceans*. Society for Promoting Christian Knowledge, London.

Gosse, P. H. (1856). *The aquarium*. Jan van Voorst, London.

Gosse, P. H., (1860). *A history of the British sea anemones and corals*. Jan van Voorst, London.

Graham, M. (1935). Modern theory of exploiting a fishery and application to North Sea trawling. *J. Cons. Int. Explor. Mer.* **10**: 264–74.

Graham, M. (1938). Rates of fishing and natural mortality from the data of marking experiments. *J. Cons. Int. Explor. Mer.* **10**: 264–74.

Graham, M. (1939). The sigmoid curve and the overfishing problem. *Rapp. Procès-Verb. Cons. Int. Explor. Mer.* **110**: 15–20.

Graham, M. (1943). *The fish gate*. Faber, London.

Graham, M. (1950). A review of the conference. In *Proceedings of UNSCURR 1949*, Lake Success, vol. I, pp. 410–11. United Nations, New York.

Gray, M. (1978). *The fishing industries of Scotland, 1790–1814*. Oxford University Press, Oxford.

Gulland, J. A. (1961a). Fishing and the stocks of fish at Iceland. *Fish. Invest. Lond.* ser. 2, **23**.

Gulland, J. A. (1961b). The estimation of the effect on catches of changes in gear selectivity. *J. Cons. Int. Explor. Mer.* **26**: 204–14.

Gulland, J. A. (1965). Estimation of mortality rates. Annex North East Arctic Working Group Int. Council Explor. Sea CM G3. Mimeo.

Gulland, J. A. (1966). The effect of regulation on Antarctic whale catches. *J. Cons. Int. Explor. Mer.* **30**: 308–15.

Gulland, J. A. (1982). Long term potential effects from management of the fish resources of the North Atlantic. *J. Cons. Int. Explor. Mer.* **40**: 8–16.

Gulland, J. A. (1983). World resources of fisheries and their management. In *Marine ecology*, vol. V, part 2, ed. O. Kinne, pp. 839–1060. John Wiley and Sons, London and New York.

Gulland, J. A. and Boerema, L. K. (1973). Scientific advice on catch levels. *Fish. Bull. US Dept Commerce Fish Wildl. Serv.* **71**: 325–35.

Hanna, G. A. (1926). Expedition to the Revillagigedo Is. Mexico in 1925. *Proc. Calif. Acad. Sci.* ser 4, **15**: 115–94.

Hardy, A. C. (1959). *Fish and Fisheries.* Collins, London.

Hardy, Sir A. (1967). *Great Waters.* Collins, London.

Herdman, W. A. (1904). *Report to the Government of Ceylon on the pearl oyster fisheries of the Gulf of Mannar*, part 11. Royal Society, London.

Heincke, F. (1913). Untersuchungen über die Scholle. Generallbericht 1. Schollenfischerei und Schonmassregeln. Verlaufige Kurze Ubersicht über wichtigsten Ergebnisse des Berichts. *Rapp. Procès-Verb. Cons. Int. Explor. Mer.* **16**: 1–70.

Helgason, T. and Gislason, H. (1979). V.P.A. analysis with species interaction due to predation. ICES CM 1979 G52. Mimeo.

Hercher, R. (ed.) (1971). *Claudius Aelianus: De Natura Animalium, Libri XVII*, vols. 1 and 2. Akademische Druk v. Verlagsanstalt, Graz.

Hérubel, M. (1912). *Sea fisheries, their treasures and toilers.* Fisher Unwin, London, Leipzig.

Hjort, J. (1933). Whales and whaling. *Hvalr. Skr.* **7**: 7–29.

Hjort, J. (1934). *The restrictive law of population.* Huxley Memorial Lecture, Imperial College, London.

Hjort, J., Jahn, G. and Ottestad, P. (1933). The optimum catch. *Havl. Skr.* **7**: 92–127.

Hodgson, W. C. (1957). *The herring and its fishery.* Routledge and Kegan Paul, London.

Höglund, H. (1972). On the Bohuslan herring during the great herring fishery period in the eighteenth century. *Rep. Inst. Mar. Res.* **20**: 1–86.

Holdsworth, E. W. H. (1874). *Deep sea fishing and fishing boats.* Edward Stanford, London.

Holt, E. W. L. (1893–5). North Sea Investigations. II. On the Iceland trawl fishery with some remarks on the history of North Sea trawling grounds. *J. Mar. Biol. Assn. UK*, **3**: 81–122.

Holt, E. W. L. (1895). An examination of the present state of the Grimsby Trawl Fishery, with especial reference to the destruction of immature fish. *J. Mar. Biol. Assn UK* N.W., **3**: 339–448.

Hornell, J. (1905). *Report to the Government of Madras on the Indian pearl fisheries in the Gulf of Mannar.* Madras.

Hornell, J. (1914). The sacred chank of India. *Madras Fish. Bur. Bull.* **7**.

Hornell, J. (1916a). An explanation of the irregularly cyclic character of the pearl fisheries of the Gulf of Mannar. *Madras Fish. Dept Bull.* **8**: 11–22.

Hornell, J. (1916b). The Indian Conch (*Turbinella pyrum* Linn.) and its relation to Hindu life and religion. In *Report to the Government of Baroda on the Marine Zoology of Okhmandel in Kattiawar*, pp. 1–78. Williams and Norgate, London.

Hornell, J. (1917). The Indian Bêche de Mer industry; its history and recent revival. Report 4, *Madras Fish. Dept Bull.* **11**: 119–50.

Hornell, J. (1925). The fishing methods of the Madras Presidency. *Madras Fish. Dept Bull.* **18**: 59–110.

Hornell, J. (1938). The fishing methods of the Madras Presidency. *Madras Fish. Dept Bull.* **27**.

Hornell, J. (1950). *Fishing in many waters*. Cambridge Univ. Press, Cambridge.

Horwood, J. W. (1986). The distribution of the Southern blue whale in relation to recent estimates of abundance. *Sci. Rep. Whal. Inst.* **37**: 135–65.

Horwood, J. W., Best, P. B. and Ohsumi, S. (1981). Southern hemisphere minke whale assessment cruise, 1979–80. *Polar Record* **20**: 565–70.

Hosius, C. (ed.) (1967). *Die Moselgedichte des Decimus Ausonius und des Venantius Fortunatus*. Georg Olms, Hildesheim.

Houghton, R. and Flatman, S. (1981). The exploitation pattern, density dependent catchability and growth of cod (*Gadus morhua*) in the west central North Sea. *J. Cons. Int. Explor. Mer.* **39**: 271–87.

Houghton, W. (1870). *Sea side walks of a naturalist with his children*. Groombridge and Sons, London.

Hourston, A. S. (1980). The decline and recovery of Canada's Pacific herring stocks. *Rapp. Procès-Verb. Cons. Int. Explor. Mer.* **177**: 143–53.

Huppert, D. D. (1985). Pacific Coast groundfish management: evolution and prospects. In *Fisheries management: issues and options*, pp. 309–35. Alaska Sea Grant Rep. 85.

Huxley, T. H. (1884). Inaugural address of the Fishery Conferences. *Fisheries Exhibition Literature* **4**: 1–19.

Iliffe Robson, E. (transl.) (1929–30). *Arrianus Flavius: Periplus of the Erythrean Sea*. Loeb Series.

Ingersoll, E. (1887). The oyster, scallop, clam, mussel and abalone industries. In *The fisheries and fishing industry of the United States*, sect. II *A grographical review of the fishing industries and fishing communities in the year 1880*, ed. G. Brown Goode, vol. 1, pp. 507–626. Govt Printing Offices, Washington, DC.

Innis, A. (1940). *The cod fisheries; the history of an international economy*. Yale Univ. Press, New Haven, Conn.

INPFC (1962). The exploitation, scientific investigation and management of halibut (*Hippoglossus stenolepis* Schmidt) stocks on the Pacific coast of North America in relation to the abstention provisions of the North Pacific Fisheries Convention. *Bull. Int. North Pac. Fish. Commn*, 7.

Jackson, G. (1978). *The British whaling trade*. Adam and Charles Black, London.

Jakobsson, J. (1980). Exploitation of the Icelandic spring and summer spawning herring in relation to fisheries management, 1947–77. *Rapp. Procès-Verb. Cons. Int. Explor. Mer.* **177**: 23–42.

Jenkins, F. (1946). *Port war*. Cowell, Ipswich and London.

Jenkins, J. T. (1921). *A history of the whale fisheries*. Constable, London.

Jenkins, J. T. (1927). *The herring and the herring fisheries*. P. S. King and Son Ltd, London.

Jones, R. (1961). The assessment of the long term effects of changes in gear selectivity and fishing effort. *Mar. Res. Scot.* **2**.

Jones, R. (1974). Assessing the long term effects of changes in fishing effort and mesh size from length composition data. I.C.E.S. C.M. 1974 F33. Mimeo.

Jones, R. (1981). The use of length composition data in fish stock assessment (with notes on V.P.A. and cohort analysis). *FAO Fish Circ.* **734**.

Jonsgård, A. (1964). A right whale (*Balaena* sp.), in all probability a Greenland right whale (*Balaena mysticetus*) observed in the Barents Sea. *Norsk Hval.* **53**: 311–13.

Jonsgård, A. (1981). Bowhead whales, *balaena mysticetus*, observed in Arctic waters of the Eastern North Atlantic after the second world war. *Rep. Int. Whaling Commn*, **31**: 511.

Jordan, D. S. and Clark, G. A. (1898). The history, conditions and needs of the herd of fur seals resorting to the Pribilov Islands. In *The fur seals and fur seal islands of the North Pacific Ocean*, ed. D. S. Jordan, L. Stejner, F. A. Lucas, J. F. Moser, C. H. Townsend, G. A.Clark and J. Murray, Part I. Govt Printing Office, Washington, DC.

Jordan, D. S. and Gilbert, C. H. (1887). The salmon fishery and the canning interests of the Pacific coast. In *The fisheries and fishing industry of the United States*, sect. I, vol. 1, pp. 729–53. Govt Printing Offices, Washington, DC.

Judah, C. B. (1933). The North American Fisheries and British policy to 1713. *Univ. Illinois Bull.* **31**.

Kashara, H. (1961). *Fisheries resources of the North Pacific Ocean*, part I, H. R. Macmillan Lectures, Univ. of British Columbia, Vancouver.

Kasahara, H. (1964). *Fisheries resources of the North Pacific Ocean*, part II, pp. 137–202. H. R. MacMillan Lectures. Univ. of British Columbia, Vancouver.

Kasahara, H. (1972). Japanese distant water fisheries: a review. *Fish. Bull. US Dept Commerce Fish Wildl. Serv.* **70**: 227–82.

Kato, H. (1983). Some considerations on the decline in age at sexual maturity of the Antarctic minke whale. *Rep. Int. Whal. Commn*, **33**: 393–9.

Kent, J., Watson, E. R. and Little, J. I. (1937). *Report of the Commission of Enquiry investigating the sea fisheries of Newfoundland and Labrador, other than the seal fishery*. Newfoundland Govt, St John's, Newfoundland.

Kenyon, K. W., Scheffer, V. B. and Chapman, D. G. (1954). A population study of the Alaskan fur seal herd. *Spec. Sci. Rep. US Dept Commerce Fish Wildl. Serv.* **12**.

Kipling, R. (1891). *The Jungle Book*. Macmillan, London.

Klawe, W. L. (1963). Observations on the spawning of four species of tuna (*Neothunnus macropterus*, *Katsuwonus pelamis*, *Auxis thazard* and *Euthynnus lineatus*) in the eastern Pacific Ocean, based on the distribution of their larvae and juveniles. *Bull. Inter-Amer. Trop. Tuna Commn*, **6**: 449–540.

Kondo, K. (1980). The recovery of the Japanese sardine – the biological basis of stock size fluctuations. *Rapp. Procès-Verb. Cons. Int. Explor. Mer.* **177**: 332–54.

Koslow, J. A. (1976). Pacific pollock – already overfished? *Sea Frontiers*, **22**: 98–105.

Kyle, H. M. (1905). Statistics of the North Sea. Part II. Summary of the available fisheries statistics and their value for the solution of the problems of overfishing. *Rapp. Procès-Verb. Cons. Int. Mer.* **3**, Appx K.

Kyle, H. M. (1929). *Handbuch der SeeFischerei Nordeuropas*, Ed. H. Lubbert and E. Ehrenbaum, Bd VI *Die Seefischerei von GrossBritannien und Irland.* Nägele, Stuttgart.

Lacépède, B. Comte de (1841). *Histoire naturelle*, Tome II *Histoire naturelle des poissons*, nouvelle édition. Furne et Cie, Paris.

Lambert, S. (1975). *House of Commons Sessional papers of the eighteenth century.* Scholarly Resources Inc., Willmington, DE.

Lander, R. H. (1975). Method of determining natural mortality in the northern fur seal (*Calliorhinus ursinus*) from known pups and kill by age and sex. *J. Fish. Res. Bd Can.* **32**: 2447–52.

Lander, R. H. (1979). Alaska or northern fur seal. *FAO Fish*, Ser. 5, **2**: 19–23.

Lander, R. H. and Kajimura, H. (1982). Status of the Northern Fur Seals. *FAO Fish*, Ser. 5, **4**: 319–46.

Larkin, P. A. and Gazey, W. (1982). Application of ecological simulation models to management of tropical multispecies fisheries. In *Proc. ICLARM/CSIRO Workshop 12–21 January 1981*, ed. D. Pauly and G. I. Murphy, pp. 123–40. ICLARM, Cronulla and Manila.

Lavigne, D. (1981). Harp seals. *FAO Fish*, Ser. 5, **2**: 76–80.

Lavigne, D. M., Innes, S., Kalpakis, K. and Ronald, K. (1975). *An aerial census of western Atlantic harp seals (*Pagophilus groenlandicus*) using ultra violet photography*. Res. Doc. 44 Int. Commn North West Atl. Fish.

Laws, R. M. (1961). Reproduction, growth and age of southern Fin whales. *Discovery Rep.* **31**: 327–486.

Laws, R. M. and Purves, P. E. (1956). The ear plug of the Mysticeti as an indiation of age with special reference to the North Atlantic fin whale (*Balaenoptera physalis* Linn.). *Norsk. Hval.* **45**: 413–25.

Le Cren, E. D. (1958). Observations on the growth of perch (*Perca fluviatilis*) over twenty two years with special reference to the effects of temperature and changes in population density. *J. Anim. Ecol.* **27**: 287–334.

Lea, E. (1929). The herring's scale as a certificate of origin. Its applicability to race investigations. *Rapp. Procès-Verb. Cons. Int. Explor. Mer.* **65**: 100–17.

Leaman, B. M. and Beamish, R. J. (1984). Ecological and management implications of longevity in some north east Pacific groundfish. *Bull. Int. North Pac. Commn*, **42**: 85–97.

Leigh, M. (1983). *European integration and the Common Fisheries Policy.* Croom Helm, London and Canberra.

Lett, P. F. and Benjaminsen, T. (1977). A stochastic model for the management of the North West Atlantic harp seal (*Pagophilus groenlandica*) population. *J. Fish Res. Bd Can.* **35**: 1155–87.

Lett, P. F., Kohler, A. C. and Fitzgerald, D. W. (1975). Role of stock biomass and temperature in recruitment of southern Gulf of St Lawrence cod, *Gadus morhua. J. Fish. Res. Bd Can.* **32**: 1613–27.

Lett, P. F., Mohn, R. K. and Gray, D. F. (1981). Density dependent processes and management strategy for the North West Atlantic Harp Seal population. In *Dynamics of large mammals*, ed. C. W. Fowler and T. D. Smith, pp. 135–57. John Wiley and Sons, New York.

Lhote, H. (1959). *The story of prehistoric rock paintings of the Sahara.* Hutchinson, London.

Lounsbury, R. G. (1934). *The British Fishery at Newfoundland, 1634–1763.* Yale Univ. Press, New Haven.

Lundberg, R. (1886). The Fisheries of Sweden. In *US Bureau of Fisheries 1884 Appendix to the Report of Commissioners*, pp. 363–98. Govt Printing Office, Washington, DC.

MacCall, A. D. (1976). Density dependence of catchability coefficient in the Californian Pacific sardine *Sardinops sagax caerulea. Rep. Calif. Coop. Ocean. Fish. Invest.* **18**: 136–48.

MacCall, A. D. (1980). Population models of the northern anchovy (*Engraulis mordax*). *Rapp. Procès-Verb. Cons. Int. Explor. Mer.* **177**: 292–306.

MacDonald, R. D. S. (1984). Canadian fisheries policy and the development of Atlantic coast groundfisheries management. In *Atlantic fisheries and coastal communications: fisheries decision making case studies*, ed. C. Lawson and A. J. Hansen, pp. 15–76. Dalhousie Ocean Studies Programme, Halifax, NS.

McFarland, R. (1911). *A history of the New England fisheries.* Appleton, New York.

McHugh, J. L. (1972). Jeffersonian democracy and the fisheries. In *World fisheries policy*, ed. B. J. Rothschild, pp. 134–57. Univ. of Washington Press, Seattle.

McHugh, J. L. (1983). Jeffersonian democracy and the fisheries revisited. In *Global fisheries*, ed. B. J. Rothschild, pp. 73–96. Springer-Verlag, Heidelberg.

McKerrow, R. B. (ed.) (1901–10). *The Works of Thomas Nashe.* A. H. Bullen, London.

McPherson, N. L. (1935). *The dried codfish industry.* Newfoundland Dept of Natural Resources, St Johns.

Mackintosh, N. A. and Wheeler, J. F. G. (1929). Southern blue and fin whales. *Discovery Rep.* **1**: 257–540.

Mair, A. W. (transl.) (1928). *Oppian: Colluthus Tryphiodorus.*

Manzer, J. I., Ishida, T., Peterson, A. E. and Hanavan, M. G. (1965). Salmon of the North Pacific Ocean. V. Offshore distribution of salmon. *Bull. Int. Pacific Fish. Commn*, **15**: 1–452.

March, E. (1952). *Sailing drifters.* David and Charles, Newton Abbott.

Margolis, L. (1963). Parasites as the geographical origin of the sockeye salmon, *Oncorhynchus nerka* Walbaum, occurring in the North Pacific Ocean and adjacent seas. *Bull. Int. North Pac. Fish. Commn*, **11**, 101–56.

Margolis, L., Cleaver, F. C., Fukuda, Y. and Godfrey, H. (1966). Salmon of the North Pacific. VI. Sockeye salmon in offshore waters. *Bull. Int. North Pacific Fish. Commn*, **20**, 1–70.

Marquette, W. M. and Bockstoce, J. R. (1980). Historical shore based catch of Bowhead whales in the Bering, Chuckchi and Beaufort Seas. *Mar. Fish. Rev.* Sept.–Oct., pp. 39–69.

Marr, J. C. (1960). The causes of major variations in the catch of Pacific sardine (*Sardinops caerulea* Girard). *World Sci. meeting on the biology of sardines*, vol. 3, pp. 1–69. FAO, Rome.

Masaki, Y. (1976). Biological studies on the North Pacific sei whale. *Bull. Far Seas Fish Res. Lab.* **14**: 1–104.

May, R. M., Beddington, J. R., Clark, C. W., Holt, S. J. and Laws, R. M. (1979). Management of multispecies fisheries. *Science*, **205**: 267–77.

Miles, E., Gibbs, S., Fluharty, D., Dawson, C. and Teeter, D. (1982). *The Management of Marine Regions: The North Pacific*. Univ. of California Press, Berkeley, CA.

Mitchell, E. (1977). Initial population size of bowhead whale (*Balaena mysticetus*) stocks: cumulative catch estimates. Unpublished paper SC 29/Doc 33 Sci. Cttee June 1977, Int. Whaling Commn.

Mitchell, E. and Reeves, R. E. (1981). Catch history and cumulative catch estimates of initial population size of cetaceans in East Canadian Arctic. *Rep. Int. Whal. Commn*, **31**, 645–82.

Mitchell, J. M. (1864). *The herring, its natural history and national importance*. Edmonston and Douglas, Edinburgh.

Morgan, G. R. (1980). Increases in fishing effort in a limited entry fishery – the Western Rock Lobster fishery, 1963–1976. *J. Cons. Int. Explor. Mer*. **39**: 82–7.

Mosdell, H. M. (1923). *Chafe's sealing books*. The Trade Printers, St John's, Newfoundland.

Murphy, G. I. (1965). A solution of the catch equation. *J. Fish. Res. Bd Can*. **22**: 191–202.

Murphy, G. I. (1977). Clupeoids. In *Fish population dynamics*, ed. J. A. Gulland, pp. 283–308. John Wiley, London.

Murray, Sir J. and Hjort, J. (1912). *The depths of the ocean*. Macmillan, London.

Nagasaki, F. (1961). *Population study on the fur seal herd*. Contrib. no. 365 Tokai Regional Fisheries Research Laboratory.

Nall, J. G. (1866). *Great Yarmouth and Lowestoft, a handbook for visitors and residents; with chapters on the archaeology, natural history etc. of the district; a history with statistics of the East coast herring fishery and an etymological and comparative glossary of the dialect of East Anglia*. Longman Green, London.

Newman, G. G. and Crawford, R. J. M. (1980). Population biology and management of mixed-species pelagic stocks off South Africa. *Rapp. Procès-Verb. Cons. Int. Explor. Mer*. **177**, 279–91.

Ohsumi, S. and Yamamura, K. (1981). A review of the Japanese whale sightings system. *Rep. Int. Whal. Commn*, **32**: 581–6.

Olsen, O. W. (1958). Hookworms, *Uncinaria lucasi* Stiles 1901, in fur seals, *Calliorhinus ursinus* Linn. on the Pribilov Islands. *Trans. North Amer. wildlife Conf*. **23**, 152–75.

O'Reilly, J. E. and Busch, D. A. (1984). Phytoplankton production on the North west Atlantic shelf. *Rapp. Procès-Verb. Cons. Int. Explor. Mer*. **183**, 255–68.

Otsu, T. (1960). Albacore migration and growth in the North Pacific Ocean. *Fish. Bull. US Dept Commerce Fish Wildl. Serv*. **63**: 33–44.

Ottestad, P. (1956). On the size of the stock of Antarctic fin whales relative to the size of the catch. *Norsk. Hval*. **45**: 298–308.

Ottestad, P. (1969). Forecasting the annual yield in sea fisheries. *Nature*, **185**: 183.

Palmer, C. J. (1854). *The History of Great Yarmouth*, ed. H. Manship, vols. 1 and 2.

Parona, C. (1919). Il Tonno e la sua Pesca. *R. Comm. Talass. Ital*. **58**: 1–265.

Pauly, D. (1979). Theory and management of tropical multispecies stocks: a review with emphasis on the South East Asian demersal fisheries. *ICLARM Studies and Reviews* 1.

Pauly, D. and David, N. (1980). An objective method for determining fish growth from length frequency data. *ICLARM Newsletter*, 3: 13–15.

Pella, J. J. and Tomlinson, P. K. (1969). A generalized stock production model. *Bull. Inter-Amer. Trop. Tuna Commn*, 13: 421–58.

Petersen, C. B. J. (1894). On the biology of our flatfishes and on the decrease of our flatfish fisheries. *Rep. Dan. Biol. Stn 1893*.

Petersen, C. G. J. (1896). The yearly immigration of young plaice into the Limfjord from the German Sea. *Rep. Dan. Biol. Stn 1895*, 6: 1–48.

Petersen, C. G. J. (1900). What is overfishing? *J. Mar. Biol. Assn UK*, 6: 587–94.

Pope, J. G. (1972). An investigation of the accuracy of virtual population analysis using cohort analysis. *Res. Bull. Int. Commn North West Atl. Fish.* 1: 163–9.

Pope, J. G. (1976). The effect of biological interaction on the theory of mixed fisheries. *Selected papers Int. Commn North West Atl. Fish.* 1: 157–62.

Pope, J. G. (1979). A modified cohort analysis in which constant natural mortality is replaced by estimates of predation levels. *I.C.E.S. CM 1979* H16. Mimeo.

Pope, J. G. (1982). Short cut formulae for the estimation of coefficients of variation of status quo T.A.C.s. I.C.E.S. CM 1982 G12. Mimeo.

Pope, J. G. and Garrod, D. J. (1975). Sources of error in catch and effort quota regulations with special reference to variations in the catchability coefficient. *Res. Bull. Int. Commn North West Atl. Fish.* 11: 17–30.

Pritchard, A. L. (1939). Homing tendency and age at maturity of pink salmon (*Oncorhyncus gorbuscha*) in British Columbia. *J. Fish Res. Bd Can.* 4: 141–50.

Prowse, D. W. (1896). *A history of Newfoundland from the English, Colonial and Foreign Records*. Eyre and Spottiswoode, London.

Purves, P. E. (1955). The wax plug in the external auditory meatus of the Mysticeti. *Discovery Rep.* 27: 293–302.

Radcliffe, W. (1921). *Fishing from the Earliest Times*. John Murray, London.

Raleigh, W. (1603). Observations concerning the Trade and Commerce of England with the Dutch and other foreign nations, MS laid before James I. In Samuel, A. M. (1918). *The herring: its effect on the history of Britain*, p. 94. John Murray, London.

Rathbun, R. (1887). The crab, lobster, crayfish, rock lobster, shrimp and prawn fisheries. In *The fisheries and fishing industry of the United States*, sect. II *A geographical review of the fishing industries and fishing communities in the year 1880*, ed. G. Brown Goode, vol. 1, pp. 629–810. Govt Printing Offices, Washington, DC.

Rau, C. (1884). *Prehistoric fishing in Europe and North America*. Smithsonian Institute, Washington, DC.

Ricker, W. E. (1944). Further notes on fishing mortality and effort. *Copeia 1944*: 23–44.

Ricker, W. E. (1946). Production and utilization of fish populations. *Ecol. Monogr.* 16: 373–91.

Ricker, W. E. (1948). Methods of estimating vital statistics of fish populations. *Indiana Univ. Publ. Sci. Ser.* 15.

Ricker, W. E. (1954). Stock and recruitment. *J. Fish. Res., Bd Can.* 11: 559–623.

Ricker, W. E. (1958). Handbook of computations for biological statistics of fish populations. *Bull. Fish. Res. Bd Can.* **119**.

Riley, F. (1967). *Fur seal industry of the Pribilov Islands.* Circ. 275 US Dept Commerce Fish and Wildl. Serv.

Roe, H. S. J. (1967). Seasonal formation of laminae in the ear plug of the fin whale. *Discovery Rep.* **35**: 1–30.

Rothschild, B. J. (1970). King crab systems study. In *Research in Fisheries*, p. 18. Univ. Wash. Coll. Fish., Seattle.

Rothschild, B. J. (1983). Achievement of fisheries management goals in the 1980s. In *Global fisheries*, ed. B. J. Rothschild, pp. 151–77. Springer-Verlag, Heidelberg.

Royce, W. F. (1964). A morphometric study of yellowfin tuna *Thunnus albacares* Bonnaterre. *Fish Bull. US Dept Commerce Fish Wildl. Serv.* **63**, 395–443.

Russell, E. S. (1914). Report on the market measurements in relation to the English haddock fishery during the years 1909–11. *Fish. Invest. Lond.* ser. 2, **1**.

Russell, E. S. (1922). Report on the market measurements in relation to the English cod fishery during the years 1912–14. *Fish. Invest. Lond.* ser. 2, **5**.

Russell, E. S. (1931). Some theoretical considerations on the overfishing problem. *J. Cons. Int. Explor. Mer.* **6**: 3–20.

Russell, E. S. (1932). Is the destruction of undersized fish by trawling prejudicial to the stock? *Rapp. Procès-Verb. Cons. Int. Explor. Mer.* **80**: 8.

Russell, E. S. (1934). Size limits and mesh regulation for sea fish. *Rapp. Procès-Verb. Cons. Int. Explor. Mer.* **105**: 1–8.

Russell, E. S. (1939). An elementary treatment of the overfishing problem. *Rapp. Procès-Verb. Cons. Int. Explor. Mer.* **110**: 5–14.

Russell, P. (1951). Some historical notes on the Brixham fisheries. *Trans. Devonshire Ass. Adv. Sci.* **83**.

Ruud, J. T. (1940). The surface structure of the baleen plates as a possible clue to age in whales. *Hval. Skr.* **29**.

Ruud, J. T. (1945). Further studies on the structure of baleen plates and their application to age determination. *Hval. Skr.* **29**.

Ruud, J. T., Jonsgaard, A. and Ottestad, P. (1950). Age studies on blue whales. *Hval. Skr.* **33**.

Sabine, L. (1853). Report on the principal fisheries of the American Seas. In *US Treasury Dept. ann. rep. of the Secretary on the state of the finances*, pp. 181–493. Washington, DC.

Samuel, A. M. (1918). *The herring: its effect on the history of Britain.* John Murray, London.

Saul, A. (1981). The herring industry at Great Yarmouth c. 1280–1400. *Norfolk Archaeology*, **38**: 33–41.

Saville, A. and Bailey, R. S. (1980). The assessment and management of the herring stocks in the North Sea and west of Scotland. *Rapp. Procès-Verb. Cons. Int. Explor. Mer.* **177**: 112–43.

Schaaf, W. E. (1980;. An analysis of the dynamic population response of Atlantic menhaden (*Brevoortia tyrannus*) to an intensive fishery. *Rapp. Procès-Verb. Cons. Int. Explor. Mer.* **177**: 243–51.

Schäfer, D. (1887). *Das Buch des Lübeckischen Vogts auf Schonen.* Hansische Geschichtsquellen 4, Halle.

Schaefer, M. B. (1954). Some aspects of the dynamics of populations important to the management of the commercial fish populations. *Fish. Bull. US Dept Int. Fish Wildl. Serv.* **52**: 191–203.

Schaefer, M. B. (1955). Morphometric comparison of yellowfin tuna from Southeast Polynesia, Central America and Hawaii. *Bull. Inter-Amer. Trop. Tuna Commn*, **1**: 91–136.

Schaefer, M. B. (1957). A study of the dynamics of the fishery for the yellowfin tuna in the eastern tropical Pacific Ocean. *Bull. Inter-Amer. Trop. Tuna Commn*, **2**: 245–85.

Schaefer, M. B. and Beverton, R. J. H. (1963). Fishery dynamics, their analysis and interpretation. In *The Sea*, ed. M. N. Hill, vol. 2, pp. 464–83. Wiley Interscience, New York.

Schaefer, M. B. and Orange, C. J. (1956). Studies of the sexual development and spawning of yellowfin tuna (*Neothunnus macropterus*) and skipjack (*Katsuwonus pelamis*) in three areas of the Pacific Ocean, by examination of gonads. *Bull. Inter-Amer. Trop. Tuna Commn*, **1**: 283–349.

Scheffer, V. B. (1950*a*) *The food of the alaska fur seal*. US Dept. Int. Wildlife leaflet no. 329.

Scheffer, V. B. (1950*b*). Growth layers on the teeth of Pinnipedia as an indication of age. *Science*, **112**: 309–11.

Schopka, S. A. (1980). The recent changes in fishing pattern in Icelandic waters and their effects on the yield of cod and haddock stocks. I.C.E.S. CM 1980 G42. Mimeo.

Schumacher, A. (1980). Review of North Atlantic catch statistics. *Rapp. Procès-Verb. Cons. Int. Explor. Mer.* **177**: 8–22.

Scoresby, W. (1820). *An account of the Arctic Regions with a history and description of the Northern Whale Fishery*, vol. 2 *The Whale Fishery*. Constable, Edinburgh.

Sergeant, D. E. (1975). Estimating numbers of harp seals. *Rapp. Procès-Verb. Cons. Int. Explor. Mer.* **169**: 274–80.

Serobaba, I. I. (1977). Data on the population structure of the Bering Sea Alaska pollock, *Theragra chalcogramma* (Pallas). *Vop. Ikhtiol.* **17**: 247–60.

Shaughnessy, P. D. and Best, P. B. (1982). A discrete population model for the South African fur seal, *Arctocephalus pusillus pusillus*. *F.A.O. Fish*, Ser. 5, **4**: 163–76.

Shepherd, J. G. (1982). A versatile new stock-recruitment relationship for fisheries and the construction of sustainable yield curves. *J. Cons. Int. Explor. Mer.* **40**: 67–75.

Shepherd J. G. (1984). Status quo catch estimation and its use in fishery management. ICES CM 1984 G5. Mimeo.

Shepherd, J. G. (1987). Towards a method for short term forecasting of catch rates based on length compositions. In *ICLARM/KISR conference on theory and applications of length-based Methods of Stock Assessment*, pp. 000–00. ICLARM/KISR. In press.

Shepherd, J.G. (1988). An exploratory method for the assessment of multispecies fisheries. *J. Cons. Int. Explor.* **44**: 189–99.

Shepherd, J. G. and Cushing, D. H. (1980). A mechanism for density dependent survival of larval fish as the basis of a stock-recruitment relationship. *J. Cons. Int. Explor. Mer.* **39**: 160–7.

Shepherd, J. G. and Stevens, S. M. (1983). Separable V.P.A.; user's guide. *Min. Agric. Fish. and Food Int. Rep.* **8**.

Sherman, K., Jones, C., Sullivan, L., Smith, W., Berrien, P. and Ejsymont, L. (1981). Congruent shifts in sandeel abundance in western and eastern North Atlantic ecosystems. *Nature* **29**: 486–8.

Sissenwine, M. P. and Marchessault, G. D. (1985). New England groundfish management: a scientific perspective on theory and reality. In *Fisheries management: issues and options.* Alaska Sea Grant Rep. **85** (2).

Sissenwine, M. P. and Shepherd, J. G. (1987). An alternative perspective on recruitment overfishing and biological reference points. *Can. J. Aquat. Sci.* In press.

Skud, B. E. (1977). Drift, migration and intermingling of Pacific halibut stocks. *Int. Pac. Halibut Commn Sci. Rep.* **63**.

Small, G. L. (1971). *The blue whale.* Columbia Univ. Press, New York.

Smith, A. D. M. and Walters, C. J. (1981). Adaptive management of stock-recruitment systems. *Can. J. Aquat. Sci.* **38**: 690–703.

Smith, J. K. (1876). Historical observations on the conditions of the fisheries among the ancient Greeks and Romans and on their mode of salting and pickling fish. *Rep. Comm. US Comm. Fish and Fisheries for 1873–4 and 1874–5*, part iii, Appx 3, pp. 3–20.

Smith, M. E. (1982). Fisheries management: intended results and unintended consequences. In *Modernization and marine fisheries policy*, ed. J. R. Maido and M. K. Orbach, pp. 57–93. Ann Arbor Science, Ann Arbor, MI.

Smith, T. and Polacheck, T. (1981). Reexamination of the life table for northern fur seals with implications about population regulating mechanisms. In *Population dynamics of large mammals*, ed. C. W. Fowler and T. D. Smith, pp. 99–120. John Wiley and Sons, New York.

Southward, G. M. (1967). Growth of Pacific halibut. *Rep. Int. N Pac. Halibut Commn.* **43**.

Sparre, P. (1980). A goal function of fisheries (Legion analysis). ICES CM 1980 G40. Mimeo.

Spence, W. (1980). *Harpooned, the story of whaling.* Conway, Greenwich.

Starbuck, A. (1878). History of the American whale fishery from its earliest inception to the year 1876. *Rep. Comm. U.S. Comm. Fish and Fisheries 1875–76*, Appx, pp. 1–769.

Stearns, S. (1887). The red snapper and Havana market fisheries. In *The fisheries and fishing industry of the United States*, sect. I *Natural history of aquatic animals*, ed. G. Brown Goode, vol. 1, pp. 585–94. Govt Printing Offices, Washington, DC.

Suzuki, Z., Tomlinson, P. K. and Honma, M. (1978). Population structure of Pacific yellowfin tuna. *Bull. Inter-Amer. Trop. Tuna Commn*, **17**: 277–358.

Swinden, H. (1772). *The history and antiquities of the ancient burgh of Great Yarmouth in the County of Norfolk.* John Crouse, Norwich.

Tanaka, S. (1962). On the salmon stocks of the Pacific coast of the United States and Canada (views of the Japanese National Section on the abstention cases of the United States and Canada). *Bull. Int. North Pacific Fish. Commn*, **9**: 69–84.

Tanaka, S. (1984). On the method of calculating catch quotas. *Bull. Int. North Pacific Fish. Commn*, **42**: 98–103.

Taylor, F. H. C., Fujinaga, M. and Wilke, F. (1955). *Distribution and food habits of the fur seals of the North Pacific Ocean*. US Dept. Int. Fish Wildl. Serv. Wilfl. Leaflet no. 329.

Templeman, W. and Gulland, J. A. (1965). Review of possible conservation actions for the I.C.N.A.F. area. *Int. Commn North West Atl. Ann. Proc.* **15**: 47–56.

Thomson, J. (1849). *The value and importance of the Scottish fisheries*. Smith Elder, Cornhill, London.

Thompson, d'Arcy W. (1909). On the statistics of the Aberdeen trawl fishery, 1901–1906, with special reference to the cod and haddock. *Rapp. Procès-Verb. Cons. Int. Explor. Mer.* **10**: B2.

Thompson, d'Arcy W. (1947). *A glossary of Greek fishes*. Oxford University Press, London.

Thompson, W. F. (1916). The problem of the halibut. *Rep. Prov. Dep. Fish. Brit. Columb. 1915*: pp. 130–40.

Thompson, W. F. (1917*a*). Statistics of the halibut fishery in the Pacific: their bearing on the biology of the species and the condition of the banks. *Rep. Commn Fish. Brit. Columb. 1916*: pp. 65–126.

Thompson, W. F. (1917*b*). The regulation of the halibut fishery of the Pacific. *Rep. Prov. Dep. Fish. Brit. Columb. 1916*: pp. 28–34.

Thompson, W. F. (1937). Theory of the effect of fishing on the stocks of halibut. *Rep. Int. Fish. Commn*, **12**.

Thompson, W. F. (1945). Effect of the obstruction at Hell's Gate on the sockeye salmon of the Fraser River. *Bull. Int. Salmon Fish. Commn*, **1**.

Thompson, W. F. (1952). Conditions of the stock of halibut in the Pacific. *J. Cons. Int. Explor. Mer.* **18**: 141–6.

Thompson, W. F. and Bell, F. H. (1934). Biological statistics of the Pacific halibut. 2. Effects of changes in intensity upon total yield and yield per unit of gear. *Rep. Int. Fish. Commn*, **8**.

Thompson, W. F. and Freeman, N. (1930). History of the halibut fishery. *Rep. Int. Fish. Commn*, **5**.

Thompson, W. F. and Herrington, W. C. (1930). Life history of the Pacific halibut. 1. Marking experiments. *Rep. Int. Fish. Commn*, **2**.

Thompson, W. F. and van Cleve, R. (1936). Life history of the Pacific halibut. 2. Distribution and early life history. *Rep. Int. Fish. Commn*, **9**.

Thursby-Pelham, D. E. (1939). The effect of fishing on the stock of plaice in the North Sea. *Rapp. Procès-Verb. Cons. Int. Explor. Mer.* **110**: 39–63.

Tillman, M. F., Breiwick, J. M. and Chapman, D. G. (1983). Reanalysis of historical whaling data for the Western Bowhead population. *Rep. Int. Whaling Commn Spec. Issue*, **5**: 143–51.

Tonnessen, J. M. and Johnsen, A. O. (1982). *The history of modern whaling*. Hurst, London.

Townsend, C. H. (1935). The distribution of certain whales as shown by logbook records of American whaleships. *Zoologica, NY*, **18**: 1–50.

Troadec, J. P., Clarke, W. G. and Gulland, J. A. (1980). A review of some pelagic fish stocks in some other areas. *Rapp. Procès-Verb. Cons. Int. Explor. Mer.* **177**: 252–77.

True, F. W. (1887). The turtle and terrapin fisheries. In *The fisheries and fishing industry of the United States*, sect. II *A geographical review of the fishing indus-*

tries and fishing communities in the year 1880, ed. G. Brown Goode, vol. 1, pp. 493–504. Govt Printing Offices, Washington, DC.

Tuck, A. and Grenier, R. (1981). A 16th century Basque whaling station in Labrador. *Sci. Amer.* Nov.: 126–36.

Ulltang, O. (1976). Catch per unit of effort in the Californian purse seine fishery for Atlanto-Scandian (Norwegian spring spawning) herring. *FAO Fish. Tech. Paper* **155**: 91–101.

Ursin, E. (1982). Multispecies fish stock and yield assessment in I.C.E.S. *Can. Spec. Publ. Fish. Aq. Nat. Sci.* **59**: 39–47.

Viale, D. (1981). Ecologie des cétacés de la Méditerranée occidentale. *FAO Fish*, ser. 5, **3**: 287–300.

von Bertalannfy, L. (1938). A quantitative theory of organic growth (inquiries on growth laws). *Human Biol.* **10**: 181–213.

Walter, G. G. (1986). A robust approach to equilibrium yield curves. *Can. J. Aq. Sci.* **43**: 1332–9.

Warner, W. W. (1984). *Distant water: the fate of the North Atlantic fishermen.* Penguin, London.

Weddell, J. (1825). *A voyage towards the South Pole performed in the years 1822–4.* Longman, London.

White, E. W. (1950). *British fishing boats and coastal craft*, part I *Historical survey.* HMSO, London.

Wilson, G. (1965). *Scottish fishing craft.* Fishing News (Books), London.

Wimpenny, R. S. (1953). *The Plaice, being the Buckland Lectures for 1949.* Arnold, London.

Winters, G. H. (1978). Production, mortality and sustainable yield of North west Atlantic harp seals (*Pagophilus groenlandicus*). *J. Fish. Res. Bd Can.* **35**: 1249–61.

Wood, W. (1911). *North Sea Fishers and Fighters.* Kegan Paul, Trench and Tribner, London.

Woodhead, P. M. J. (1964). The death of North Sea fish during the winter of 1962–3, particularly with reference to the sole, *Solea vulgaris. Helgol. Wiss. Meeresunters*, **10**: 283–300.

Yearbooks of Fishery Statistics. FAO, Rome.

INDEX